SPLINE ANALYSIS

Prentice-Hall
Series in Automatic Computation
George Forsythe, editor

AHO AND ULLMAN, *Theory of Parsing, Translation, and Compiling,* Volume I: *Parsing;* Volume II: *Compiling*
(ANDREE)3, *Computer Programming: Techniques, Analysis, and Mathematics*
ANSELONE, *Collectively Compact Operator Approximation Theory and Applications to Integral Equations*
ARBIB, *Theories of Abstract Automata*
BATES AND DOUGLAS, *Programming Language/One*, 2nd ed.
BLUMENTHAL, *Management Information Systems*
BOBROW AND SCHWARTZ, *Computers and the Policy-Making Community*
BOWLES, editor, *Computers in Humanistic Research*
BRENT, *Algorithms for Minimization without Derivatives*
CESCHINO AND KUNTZMAN, *Numerical Solution of Initial Value Problems*
CRESS, et. al., *FORTRAN IV with WATFOR and WATFIV*
DANIEL, *The Approximate Minimization of Functionals*
DESMONDE, *A Conversational Graphic Data Processing System*
DESMONDE, *Computers and Their Uses*, 2nd ed.
DESMONDE, *Real-Time Data Processing Systems*
DRUMMOND, *Evaluation and Measurement Techniques for Digital Computer Systems*
EVANS, et al., *Simulation Using Digital Computers*
FIKE, *Computer Evaluation of Mathematical Functions*
FIKE, *PL/1 for Scientific Programers*
FORSYTHE AND MOLER, *Computer Solution of Linear Algebraic Systems*
GAUTHIER AND PONTO, *Designing Systems Programs*
GEAR, *Numerical Initial Value Problems in Ordinary Differential Equations*
GOLDEN, *FORTRAN IV Programming and Computing*
GOLDEN AND LEICHUS, *IBM/360 Programming and Computing*
GORDON, *System Simulation*
GREENSPAN, *Lectures on the Numerical Solution of Linear, Singular, and Nonlinear Differential Equations*
GRUENBERGER, editor, *Computers and Communications*
GRUENBERGER, editor, *Critical Factors in Data Management*
GRUENBERGER, editor, *Expanding Use of Computers in the 70's*
GRUENBERGER, editor, *Fourth Generation Computers*
HARTMANIS AND STEARNS, *Algebraic Structure Theory of Sequential Machines*
HULL, *Introduction to Computing*
JACOBY, et al., *Iterative Methods for Nonlinear Optimization Problems*
JOHNSON, *System Structure in Data, Programs and Computers*
KANTER, *The Computer and the Executive*
KIVIAT, et al., *The SIMSCRIPT II Programming Language*
LORIN, *Parallelism in Hardware and Software: Real and Apparent Concurrency*
LOUDEN AND LEDIN, *Programming the IBM 1130*, 2nd ed.
MARTIN, *Design of Real-Time Computer Systems*
MARTIN, *Future Developments in Telecommunications*
MARTIN, *Man-Computer Dialogue*
MARTIN, *Programming Real-Time Computing Systems*
MARTIN, *Systems Analysis for Data Transmission*

MARTIN, *Telecommunications and the Computer*
MARTIN, *Teleprocessing Network Organization*
MARTIN AND NORMAN, *The Computerized Society*
MATHISON AND WALKER, *Computers and Telecommunications: Issues in Public Policy*
MCKEEMAN, et. al., *A Compiler Generator*
MEYERS, *Time-Sharing Computation in the Social Sciences*
MINSKY, *Computation: Finite and Infinite Machines*
MOORE, *Interval Analysis*
PLANE AND MCMILLAN, *Discrete Optimization: Integer Programming and Network Analysis for Management Decisions*
PRITSKER AND KIVIAT, *Simulation with GASP II: a FORTRAN-Based Simulation Language*
PYLYSHYN, editor, *Perspectives on the Computer Revolution*
RICH, *Internal Sorting Methods: Illustrated with PL/1 Program*
RUSTIN, editor, *Algorithm Specification*
RUSTIN, editor, *Computer Networks*
RUSTIN, editor, *Data Base Systems*
RUSTIN, editor, *Debugging Techniques in Large Systems*
RUSTIN, editor, *Design and Optimization of Compilers*
RUSTIN, editor, *Formal Semantics of Programming Languages*
SACKMAN AND CITRENBAUM, editors, *On-Line Planning: Towards Creative Problem-Solving*
SALTON, editor, *The SMART Retrieval System: Experiments in Automatic Document Processing*
SAMMET, *Programming Languages: History and Fundamentals*
SCHULTZ, *Digital Processing: A System Orientation*
SCHULTZ, *Finite Element Analysis*
SCHWARZ, et al., *Numerical Analysis of Symmetric Matrices*
SHERMAN, *Techniques in Computer Programming*
SIMON AND SIKLOSSY, *Representation and Meaning: Experiments with Information Processing Systems*
SNYDER, *Chebyshev Methods in Numerical Approximation*
STERLING AND POLLACK, *Introduction to Statistical Data Processing*
STOUTMEYER, *PL/1 Programming for Engineering and Science*
STROUD, *Approximate Calculation of Multiple Integrals*
STROUD AND SECREST, *Gaussian Quadrature Formulas*
TAVISS, editor, *The Computer Impact*
TRAUB, *Iterative Methods for the Solution of Polynomial Equations*
UHR, *Pattern Recognition, Learning, and Thought*
VAN TASSEL, *Computer Security Management*
VARGA, *Matrix Iterative Analysis*
VAZSONYI, *Problem Solving by Digital Computers with PL/1 Programming*
WAITE, *Implemening Software for Non-Numeric Application*
WILKINSON, *Rounding Errors in Algebraic Processes*
WIRTH, *Systematic Programming: An Introduction*
ZIEGLER, *Time-Sharing Data Processing Systems*

SPLINE ANALYSIS

MARTIN H. SCHULTZ

Department of Computer Science
Yale University

PRENTICE-HALL, INC.

ENGLEWOOD CLIFFS, N.J.

10 9 8 7 6 5 4 3 2 1

ISBN.: 0-13-835405-7
Library of Congress Catalog Card Number: 72-1402

Printed in the United States of America

PRENTICE-HALL INTERNATIONAL, INC., *London*
PRENTICE-HALL OF AUSTRALIA, PTY. LTD., *Sydney*
PRENTICE-HALL OF CANADA, LTD., *Toronto*
PRENTICE-HALL OF INDIA PRIVATE LIMITED, *New Delhi*
PRENTICE-HALL OF JAPAN, INC., *Tokyo*

To Beverly

CONTENTS

PREFACE xiii

1 INTRODUCTION 1

Exercises 8
References 9

2 PIECEWISE LINEAR INTERPOLATION 10

2.1 One-Dimensional Problems 10
2.2 Two-Dimensional Problems 13
2.3 Error Analysis 14
Exercises 21
References 23

3 PIECEWISE CUBIC HERMITE INTERPOLATION 24

3.1 One-Dimensional Problems 24
3.2 Two-Dimensional Problems 29
3.3 Error Analysis 31
Exercises 39
References 43

4 CUBIC SPLINE INTERPOLATION 44

4.1 One-Dimensional Problems 44
4.2 Two-Dimensional Problems 49
4.3 Error Analysis 51
Exercises 61
References 63

5 LINEAR INTEGRAL EQUATIONS 64

Exercises 68
References 68

6 FINITE-ELEMENT REGRESSION 69

6.1 One-Dimensional Problems 69
6.2 Two-Dimensional Problems 76
6.3 Error Analysis 78
Exercises 84
References 86

7 THE RAYLEIGH–RITZ–GALERKIN PROCEDURE FOR ELLIPTIC DIFFERENTIAL EQUATIONS 87

7.1 Introduction 87
7.2 Linear Second-Order Two-Point Boundary Value Problems 89
7.3 Semilinear Second-Order Two-Point Boundary Value Problems 93
7.4 Second-Order Problems in the Plane 99
7.5 Error Analysis 102
Exercises 110
References 113

8 THE RAYLEIGH–RITZ–GALERKIN PROCEDURE FOR EIGENVALUE PROBLEMS 116

8.1 Introduction 116
8.2 One-Dimensional Problems 116
8.3 Error Analysis 120
Exercises 124
References 125

9 SEMI-DISCRETE GALERKIN PROCEDURES FOR PARABOLIC EQUATIONS 127

9.1 Linear Problems 127
9.2 Semilinear Problems 131
9.3 Computational Considerations 134
Exercises 139
References 140

10 THE RITZ PROCEDURE FOR AN OPTIMAL CONTROL PROBLEM 141

10.1 Formulation of the Procedure 141
10.2 Error Bounds 145
Exercises 150
References 151

INDEX 153

PREFACE

Many computational methods for solving continuous or infinite-dimensional problems with a high-speed digital computer exist. One of these, the finite-element method, seems to be almost universally applicable.

In this book I discuss, in elementary terms, a unified and mathematically rigorous approach to the finite-element method. My primary aim is to present enough practical and theoretical details to enable the reader either to implement the method intelligently on a digital computer in order to solve practical problems, or to pursue theoretical studies knowledgeably. Included in the presentation are applications to interpolation problems, integral equations, least squares or regression problems, elliptic differential equations, eigenvalue problems, parabolic problems, and optimal control problems.

The material in this book is aimed at an audience with a knowledge of calculus and linear algebra. The book can be used as a supplement to a survey text in a numerical analysis or methods course, or as a text in a finite-element methods course.

I wish to thank both the Office of Naval Research and the Chevron Oil Field Research Company for their support during the preparation of this book, Professors Stanley C. Eisenstat, Herbert Keller, John Todd, and Olof Widlund for reviewing the entire manuscript carefully, and Allon Gillon and Paul Patent for their aid with the computations. Also, I thank the Department of Computer Science at Yale University, where most of the work was completed, and Janet Gail Beyer and Vianne Ramirez, who efficiently typed the manuscript.

MARTIN H. SCHULTZ

New Haven, Connecticut

SPLINE ANALYSIS

1 INTRODUCTION

In this book, we give a unified treatment of a variety of basic numerical analysis problems. In particular, we will be concerned mainly with the question of changing infinite-dimensional or continuous problems into "discrete" ones, to obtain computationally attractive, approximate, finite-dimensional problems.

Our approach is to use variational formulations and spaces of piecewise polynomial functions. This combination was first used by Courant to study vibration problems in 1943 (cf. [1.3]) and has since been successfully adopted by engineers, who call it the "finite element procedure."

Since we wish to compute the solutions of the finite-dimensional problems on a digital computer, we are naturally led to use spaces of piecewise polynomial functions, e.g., spline functions, for which we can easily a priori construct appropriate basis functions. When coupled with the variational approach, these basis functions yield approximate, finite-dimensional problems involving sparse, well-conditioned linear systems, which can be effectively solved either by Gaussian elimination or iterative methods. Moreover, we can show that the approximate problems generally have unique solutions, and we can give general a priori error bounds, which show that the approximations obtained are high-order accurate.

In summary, we formulate a large variety of problems as minimizing a real-valued functional, F, over an infinite-dimensional function space V, and obtain computationally attractive, approximate, finite-dimensional problems by minimizing F over finite-dimensional subspaces S of V, consisting of appropriately chosen piecewise polynomial functions. Moreover, our goal is to give only a general survey of the basic ideas and general techniques of analysis and results. We will *not* attempt to present and prove the sharpest or the most general possible mathematical theorems. Instead, we will consider a variety of simple model problems. For example, we consider only second-order differential equations, domains which are either an interval or a square, and piecewise polynomials of degree one or three. We leave the extensions and generalizations to the exercises and the references, which have been chosen for their suitability in guiding the reader in further study.

We now introduce some basic mathematical notations and results, which we will use repeatedly throughout this book. We will let

$$I \equiv [0, 1] \equiv \{x \mid 0 \leq x \leq 1\},$$
$$U \equiv [0, 1] \times [0, 1] \equiv \{(x, y) \mid 0 \leq x \leq 1 \text{ and } 0 \leq y \leq 1\},$$

and for each positive integer t,

$$D^t\phi(x) \equiv \frac{d^t\phi}{dx^t}(x), \qquad D^t_x\phi(x, y) \equiv \frac{\partial^t\phi}{\partial x^t}(x, y), \qquad D^t_y\phi(x, y) \equiv \frac{\partial^t\phi}{\partial y^t}(x, y),$$

and

$$R^t \equiv \{(x_1, \ldots, x_t) \mid x_i \text{ is a real number}, 1 \leq i \leq t\},$$

i.e., R^t is Euclidean t-space. For each nonnegative integer t and for each p, $1 \leq p \leq \infty$, we will let $PC^{t,p}(a, b)$ be the set of all real-valued functions $\phi(x)$ such that:

(1) $\phi(x)$ is $t - 1$ times continuously differentiable,
(2) there exist γ_i, $0 \leq i \leq s$, with

$$a = \gamma_0 < \gamma_1 < \ldots < \gamma_s < \gamma_{s+1} = b$$

such that on each open subinterval (γ_i, γ_{i+1}), $0 \leq i \leq s$, $D^{t-1}\phi$ is continuously differentiable, and
(3) the L^p-norm of $D^t\phi$ is finite, i.e.,

$$\| D^t\phi \|_p \equiv \left(\sum_{i=0}^{s} \int_{\gamma_i}^{\gamma_{i+1}} | D^t\phi(x) |^p \, dx \right)^{1/p} < \infty.$$

For the special case of $p = \infty$, we will demand that

$$\| D^t\phi \|_\infty \equiv \max_{0 \leq i \leq s} \sup_{x \in (\gamma_i, \gamma_{i+1})} | D^t\phi(x) | < \infty.$$

Unless we state otherwise, the L^p-norm of a function ϕ of one variable, $\| \phi \|_p$, will mean the L^p-norm over $I = [0, 1]$. Similarly, for each nonnegative integer t and for each p, $1 \leq p \leq \infty$, we will let $PC^{t,p}(U)$ be the set of all real-valued functions $\phi(x, y)$ such that:

(1) $\phi(x, y)$ is $t - 1$ times continuously differentiable, i.e.,

$$D_x^l D_y^k \phi(x, y), \qquad 0 \leq l + k \leq t - 1.$$

exists and is continuous,

(2) there exist γ_i, $0 \leq i \leq s$, and μ_j, $0 \leq j \leq r$, with

$$0 = \gamma_0 < \ldots < \gamma_{s+1} = 1 \quad \text{and} \quad 0 = \mu_0 < \ldots < \mu_{r+1} = 1$$

such that on each open subrectangle,

$$(\gamma_i, \gamma_{i+1}) \times (\mu_j, \mu_{j+1}), \qquad 0 \leq i \leq s, \ \ 0 \leq j \leq r,$$

we have

$$D_x^l D_y^k \phi, \qquad 0 \leq l + k \leq t - 1,$$

continuously differentiable, and

(3) for all $0 \leq l + k \leq t$, the L^p-norm of $D_x^l D_y^k$ is finite, i.e.,

$$\| D_x^l D_y^k \phi \|_p \equiv \left(\sum_{i=0}^{s} \sum_{j=0}^{r} \int_{\gamma_i}^{\gamma_{i+1}} \int_{\mu_j}^{\mu_{j+1}} | D_x^l D_y^k \phi |^p \, dy \, dx \right)^{1/p} < \infty.$$

For the special case of $p = \infty$, we will demand that

$$\| D_x^l D_y^k \phi \|_\infty \equiv \max_{\substack{0 \leq i \leq s \\ 0 \leq j \leq r}} \sup_{(x, y) \in (\gamma_i, \gamma_{i+1}) \times (\mu_j, \mu_{j+1})} | D_x^l D_y^k \phi(x, y) | < \infty.$$

Unless we state otherwise, the L^p-norm of a function ϕ of two variables, $\| \phi \|_p$, will mean the L^p-norm over $U = [0, 1] \times [0, 1]$. Moreover,

$$PC_0^{1,p}(a, b) \equiv \{ \phi \in PC^{1,2}(a, b) \,|\, \phi(a) = \phi(b) = 0 \}$$

and

$$PC_0^{1,p}(U) \equiv \{ \phi \in PC^{1,p}(U) \,|\, \phi(x, y) = 0 \text{ for all } (x, y) \text{ in the boundary of } U, \text{ i.e., for } (x, y) \text{ with } \ x = 0 \text{ or } 1, \ \text{ or } \ y = 0 \text{ or } 1 \}.$$

Finally, we will let $\Delta : 0 = x_0 < x_1 < \ldots < x_{N+1} = 1$ be a general partition of I, with the points x_i, $0 \leq i \leq N + 1$, being called partition points or mesh points or knots, and if $\Delta_y : 0 = y_0 < y_1 < \ldots < y_{M+1} = 1$ is another such partition, we will let $\rho \equiv \Delta \times \Delta_y$ be a partition of U, i.e., ρ consists of the subrectangles of the form

$$(x_i, x_{i+1}) \times (y_j, y_{j+1}), \qquad 0 \leq i \leq N, \ \ 0 \leq j \leq M.$$

Moreover, we will let

$$h \equiv \max_{0 \le i \le N} (x_{i+1} - x_i) \quad \text{and} \quad \underline{h} \equiv \min_{0 \le i \le N} (x_{i+1} - x_i)$$

be respectively the maximum and minimum mesh lengths of Δ,

$$k \equiv \max_{0 \le j \le M} (y_{j+1} - y_j) \quad \text{and} \quad \underline{k} \equiv \min_{0 \le j \le M} (y_{j+1} - y_j)$$

be respectively the maximum and minimum mesh lengths of Δ_y, and

$$\bar{\rho} \equiv \max(h, k), \qquad \underline{\rho} \equiv \min(\underline{h}, \underline{k}).$$

We now discuss some basic mathematical results which will be used repeatedly throughout this book. We start with a generalization of Rolle's Theorem. The proof we give follows [1.4].

THEOREM 1.1

If $f \in C^n[a, b]$, $n \ge 1$, i.e., f is n times continuously differentiable on $[a, b]$, and if f has a zero of order at least m_i at x_i, $1 \le i \le k$, where

$$a = x_1 < x_2 < \ldots < x_k = b \quad \text{and} \quad \sum_{i=1}^{k} m_i \ge n + 1,$$

then there exists $\xi \in [a, b]$ such that $D^n f(\xi) = 0$. Moreover, $\xi \in (a, b)$ unless $k = 1$ and $x_1 = a$ or b, in which case $\xi = x_1$.

Proof. If $k = 1$, then the result follows by choosing $\xi = x_1$. If $k > 1$ we use induction on n.

For $n = 1$, $f(x)$ has zeroes at two distinct points and the result is just the standard Rolle's Theorem of calculus. We assume the result holds for all integers through $n - 1$ and let $g(x) = Df(x)$. The function $g(x)$ has a zero of order $m_i - 1$ at each x_i, $1 \le i \le k$, and (by the result for $n = 1$) a zero ξ_i, of order one, between each pair x_i and x_{i+1} for $1 \le i \le k - 1$. Therefore the total number of zeroes of g is

$$\sum_{i=1}^{k} (m_i - 1) + k - 1 \ge n,$$

and by the induction hypothesis there exists $\xi \in (a, b)$ such that $D^{n-1}g(\xi) = D^n f(\xi) = 1$. Q.E.D.

The next result which we will prove is called the Rayleigh–Ritz Inequality. We give the Fourier series proof due to Hurwitz; cf. [1.1].

THEOREM 1.2

If $f \in PC_0^{1,2}(a, b)$, then

$$\pi^2 \int_a^b f^2(x)\,dx \leq (b-a)^2 \int_a^b (Df(x))^2\,dx. \tag{1.1}$$

Moreover, we have equality if and only if

$$f(x) = a_1 \sin(\pi(b-a)^{-1}(x-a))$$

for some real number a_1.

Proof. Expanding $f(x)$ and $Df(x)$ in their respective Fourier series, we have

$$f(x) \sim \sum_{n=1}^{\infty} a_n \sin(n\pi(b-a)^{-1}(x-a))$$

and

$$Df(x) \sim \sum_{n=1}^{\infty} a_n n\pi(b-a)^{-1} \cos(n\pi(b-a)^{-1}(x-a)).$$

By Parseval's relation (cf. [1.1]),

$$\int_a^b (f(x))^2\,dx = \sum_{n=1}^{\infty} (a_n)^2$$

and

$$\int_a^b (Df(x))^2\,dx = \sum_{n=1}^{\infty} (a_n n\pi(b-a)^{-1})^2,$$

which implies (1.1). Q.E.D.

We come now to a discussion of the Peano Kernel Theorem. We will call E a linear functional on the vector space

$$PC^{n+1,1}(a, b), \qquad n \geq 0,$$

if E is a real-valued function on $PC^{n+1,1}(a, b)$ such that

$$E(cf) = cE(f) \quad \text{and} \quad E(f+g) = E(f) + E(g)$$

for all $f, g \in PC^{n+1,1}(a, b)$.

THEOREM 1.3

If E is a linear functional on $PC^{n+1,1}(a, b)$, $n \geq 0$, and $E(p(x)) = 0$ for all polynomials p of degree n, then for all $f \in PC^{n+1,1}(a, b)$,

$$E(f) = \frac{1}{n!}E_x\left[\int_a^b D^{n+1}f(t)(x-t)_+^n\,dt\right], \tag{1.2}$$

where

$$(x-t)_+^n \equiv \begin{cases}(x-t)^n, & x \geq t,\\ 0, & x < t,\end{cases}$$

and E_x means the linear functional E applied to the expression

$$\int_a^b D^{n+1}f(t)(x-t)_+^n\,dt,$$

considered as a function of x.

Proof. With the notations introduced, Taylor's Theorem with exact remainder can be written as

$$f(x) = f(a) + Df(a)(x-a) + \cdots + \frac{1}{n!}D^nf(a)(x-a)^n + \frac{1}{n!}\int_a^b D^{n+1}f(t)(x-t)_+^n\,dt, \tag{1.3}$$

and the result follows by applying E to both sides of the identity (1.3) and by using the linearity of E and the fact that E vanishes on all polynomials of degree n. Q.E.D.

Our next two results in this chapter will be basic to the results of Chapters 2–4.

THEOREM 1.4

If f and $g \in PC^{0,2}(I)$ and

$$(f, g)_2 \equiv \int_0^1 f(x)g(x)\,dx = 0,$$

then

$$\|f\|_2^2 + \|g\|_2^2 = \|f+g\|_2^2. \tag{1.4}$$

Proof. By definition

$$\|f+g\|_2^2 = (f,f)_2 + 2(f,g)_2 + (g,g)_2,$$

and the result follows from the orthogonality condition. Q.E.D.

COROLLARY

If f and $g \in PC^{k,2}(I)$, $k \geq 0$, and $(D^kf, D^kg)_2 = 0$, then

$$\|D^kf\|_2^2 + \|D^kg\|_2^2 = \|D^kf + D^kg\|_2^2. \tag{1.5}$$

Moreover, if $g(x)$ vanishes $n \geq k$ times on I (counting multiplicities) and $\| D^k g \|_2 = 0$, $g(x) \equiv 0$ on I.

Proof. The equality comes from Theorem 1.4 with $D^k f$ replacing f and $D^k g$ replacing g. Furthermore, by Rolle's Theorem, there exists a point $\xi \in I$ such that $D^{k-1}g(\xi) = 0$.

Hence, for all $x \in I$, we have by the Cauchy–Schwarz Inequality that

$$|D^{k-1}g(x)| = \left| \int_\xi^x D^k g(s)\, ds \right| \leq \int_\xi^x |D^k g(s)|\, ds$$
$$\leq \int_0^1 |D^k g(s)|\, ds \leq \| D^k g \|_2 = 0,$$

and g must be a polynomial of degree at most $k - 2$. But g vanishes at n points and hence must be the zero polynomial. Q.E.D.

Our last result of this chapter is called the Schmidt Inequality.

THEOREM 1.5

If $p_n(x)$ is a polynomial of degree $n = 1, 2$, or 3, then

$$\int_a^b [Dp_n(x)]^2\, dx \leq 4k_n(b - a)^{-2} \int_a^b [p_n(x)]^2\, dx, \tag{1.6}$$

where $k_1 \equiv 3$, $k_2 \equiv 15$, and $k_3 \equiv \frac{1}{2}(45 + \sqrt{1605}) \approx 42.6$.

Proof. We first consider the special case of $a = -1$ and $b = 1$. If we define the Legendre polynomials $L_0(x) \equiv \sqrt{1/2}$, $L_1(x) \equiv \sqrt{3/2}x$, $L_2(x) \equiv \sqrt{5/8}(3x^2 - 1)$, and $L_3(x) \equiv \sqrt{7/8}(5x^3 - 3x)$, then

$$\int_{-1}^1 L_i(x) L_j(x) = \delta_{ij} \equiv \begin{cases} 1, & i = j \\ 0, & i \neq j \end{cases},$$

$0 \leq i, j \leq 3$, and there exist real numbers β_0, β_1, β_2, and β_3 such that

$$p_n(x) = \sum_{i=0}^n \beta_i L_i(x). \tag{1.7}$$

Using the representation (1.7) of $p_n(x)$ in terms of Legendre polynomials, we have

$$k_n = \sup_{\boldsymbol{\beta} \neq 0} \frac{\int_{-1}^1 [Dp_n(x)]^2\, dx}{\int_{-1}^1 [p_n(x)]^2\, dx}$$
$$= \sup_{\boldsymbol{\beta} \neq 0} \frac{\boldsymbol{\beta}^T A \boldsymbol{\beta}}{\boldsymbol{\beta}^T \boldsymbol{\beta}} \equiv \sup_{\boldsymbol{\beta} \neq 0} R[\boldsymbol{\beta}],$$

where

$$A_n \equiv [a_{ij}]_{0 \le i,\, j \le n} \equiv \left[\int_{-1}^{1} DL_i(x) DL_j(x)\, dx\right]_{0 \le i,\, j \le n}$$

is symmetric, nonnegative definite and $R[\boldsymbol{\beta}]$ is the Rayleigh quotient of A_n. Furthermore, we can compute that

$$A_1 = \begin{bmatrix} 0 & 0 \\ 0 & 3 \end{bmatrix}, \quad A_2 = \begin{bmatrix} 0 & 0 & 0 \\ 0 & 3 & 0 \\ 0 & 0 & 15 \end{bmatrix}, \quad \text{and} \quad A_3 = \begin{bmatrix} 0 & 0 & 0 & 0 \\ 0 & 3 & 0 & \sqrt{21} \\ 0 & 0 & 15 & 0 \\ 0 & \sqrt{21} & 0 & 42 \end{bmatrix}.$$

By the variational characterization of the eigenvalues of symmetric matrices in terms of the Rayleigh quotient (cf. [1.2]) we have that $k_n =$ maximum eigenvalue of A_n and the inequality (1.6) follows by direct computation. To prove the inequality (1.6) for arbitrary a and b, we use the change of independent variable

$$y \equiv 2(a-b)^{-1}(a-x) - 1$$

and obtain

$$\begin{aligned}
\int_a^b [D_x p_n(x)]^2\, dx &= 2(b-a)^{-1} \int_{-1}^{1} [D_y p_n(a + \tfrac{1}{2}(y+1)(b-a))]^2\, dy \\
&\le 2(b-a)^{-1} k_n \int_{-1}^{1} [p_n(a + \tfrac{1}{2}(y+1)(b-a))]^2\, dy \\
&\le 4(b-a)^{-2} k_n \int_a^b [p_n(x)]^2\, dx.
\end{aligned}$$

Q.E.D.

EXERCISES FOR CHAPTER 1

(1.1) Show that if $w \in PC^{1,2}(a, b)$ and $w(a) = w(b) = 0$, then $\max_{a \le x \le b} |w(x)| \le \frac{1}{2}(b-a)^{1/2}\left(\int_a^b [Dw(t)]^2\, dt\right)^{1/2}$. We remark that this is a special case of the general Sobolev Inequality. (Hint: use the equality $w(x) = \frac{1}{2}\left\{\int_a^x Dw(t)\, dt - \int_x^b Dw(t)\, dt\right\}$ and the Cauchy–Schwarz Inequality.)

(1.2) Show that if $w \in PC^{1,q}(a, b)$ and $w(a) = w(b) = 0$, then $\left(\int_a^b [w(x)]^p\right)^{1/p} \le \frac{1}{2}(b-a)^{1+p^{-1}-q^{-1}}\left(\int_a^b [Dw(x)]^q\right)^{1/q}$. (Hint: Proceed as in (1.1) except using Hölder's Inequality.)

(1.3) Prove that if $f \in PC^{1,2}(a, b)$ and either $f(a) = 0$ or $f(b) = 0$, then

$$\pi^2 \int_a^b (f(x))^2 \, dx \leq 4(b-a)^2 \int_a^b (Df(x))^2 \, dx.$$

(1.4) Show that if $p \geq q$ and $\{a_i\}_{i=1}^N$ and $\{b_i\}_{i=1}^N$ are any positive numbers such that $a_i^{1/p} \leq b_i^{1/q}$, $1 \leq i \leq N$, then

$$\left(\sum_{i=1}^N a_i\right)^{1/p} \leq \left(\sum_{i=1}^N b_i\right)^{1/q}.$$

(Hint: Use Jensen's Inequality; cf. [1.1, p. 18].)

REFERENCES FOR CHAPTER 1

[1.1] BECKENBACH, E. F., and R. BELLMAN, *Inequalities*. Springer-Verlag, Berlin (1965).

[1.2] BELLMAN, R., *Introduction to Matrix Theory*. McGraw-Hill, New York (1960).

[1.3] COURANT, R., Variational methods for the solution of problems of equilibrium and vibrations. *Bull. Amer. Math. Soc.* **49**, 1–23 (1943).

[1.4] WENDROFF, B., *Theoretical Numerical Analysis*. Academic Press, New York (1966).

2 PIECEWISE LINEAR INTERPOLATION

2.1 ONE-DIMENSIONAL PROBLEMS

In this and the following two chapters, we consider interpolation procedures. Given $\Delta : 0 = x_0 < x_1 < \ldots < x_N < x_{N+1} = 1$ and $N + 2$ real numbers, $\{f_i\}_{i=0}^{N+1}$, an interpolating procedure yields a function, $g(x)$, such that $g(x)$ is defined for all $x \in I$ and $g(x_i) = f_i$, $0 \leq i \leq N + 1$. A good procedure is one which yields a function, g, which is "inexpensive" to evaluate and such that if $f_i = f(x_i)$, $0 \leq i \leq N + 1$, where $f(x)$ is a smooth function, then $g(x)$ is a good approximation to $f(x)$.

A classic procedure due to Lagrange is to let $g(x)$ be the unique N-th degree polynomial, $p_N(x)$, defined by the interpolation conditions. That is

$$p_N(x) \equiv \sum_{i=0}^{N+1} f_i l_i(x), \tag{2.1}$$

where

$$l_i(x) \equiv \prod_{\substack{j=0 \\ j \neq i}}^{N+1} (x - x_j)(x_i - x_j)^{-1}. \tag{2.2}$$

However, it is a well-known result of Runge that the Lagrange interpolation procedure is *not* good. In fact, there exist analytic functions on I, e.g., $f(x) = [(10x - 5)^2 + 1]^{-1}$, for which the sequence, $\{p_N(x)\}_{N=1}^{\infty}$, of Lagrange interpolating polynomials, defined with respect to uniform meshes, diverges; cf. [2.3].

In order to overcome this difficulty, we introduce and study a piecewise linear interpolation procedure. We begin with two basic definitions.

DEFINITION 2.1

Given Δ, let $L(\Delta)$ be the vector space of all continuous, piecewise linear polynomials with respect to Δ, i.e.,

$$L(\Delta) \equiv \{p(x) \in C(I) \mid p(x) \text{ is a linear polynomial on each subinterval } [x_i, x_{i+1}],\ 0 \leq i \leq N, \text{ defined by } \Delta\}.$$

The functions in $L(\Delta)$ are sometimes called "linear finite element functions" or "linear patch functions."

DEFINITION 2.2

Given $\mathbf{f} \equiv (f_0, f_1, \ldots, f_{N+1}) \in R^{N+2}$, let $\vartheta_{L(\Delta)}\mathbf{f}$, the *$L(\Delta)$-interpolate of* $\mathbf{f}$, be the unique element, $l(x)$, in $L(\Delta)$ such that $l(x_i) = f_i$, $0 \leq i \leq N+1$.

This procedure is well-defined. In fact, on each subinterval $[x_i, x_{i+1}]$, $0 \leq i \leq N$, $\vartheta_{L(\Delta)}\mathbf{f}$ is equal to a linear polynomial $a_i x + b_i$ which must be determined by the two conditions

$$a_i x_i + b_i = f_i \quad \text{and} \quad a_i x_{i+1} + b_i = f_{i+1}. \tag{2.3}$$

To show that there is a unique solution to (2.3), it suffices to show that the homogeneous case of $f_i = f_{i+1} = 0$ has only the solution of $a_i = b_i = 0$. However, this is obvious since $x_i \neq x_{i+1}$.

We can express the $L(\Delta)$-interpolate as

$$\vartheta_{L(\Delta)}\mathbf{f} = \sum_{i=0}^{N+1} f_i l_i(x),$$

where $l_i(x)$ is the unique element in $L(\Delta)$ such that

$$l_i(x_j) = \delta_{ij}, \qquad 0 \leq i, j \leq N+1$$

(δ_{ij} is the Kronecker delta function). The graph of $l_i(x)$, $1 \leq i \leq N$, is given by

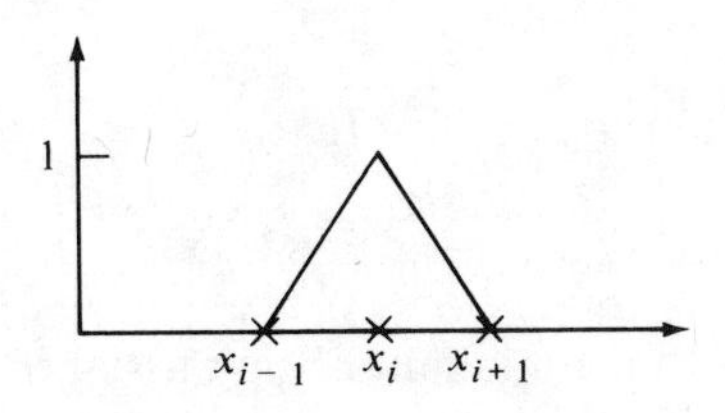

Moreover,

$$l_0(x) \equiv \begin{cases} (x_1 - x)x_1^{-1}, & 0 \le x \le x_1, \text{ and} \\ 0, & x_1 \le x \le 1, \end{cases}$$

$$l_i(x) \equiv \begin{cases} (x - x_{i-1})(x_i - x_{i-1})^{-1}, & x_{i-1} \le x \le x_i, \\ (x_{i+1} - x)(x_{i+1} - x_i)^{-1}, & x_i \le x \le x_{i+1}, \text{ and} \\ 0, & 0 \le x \le x_{i-1} \text{ or } x_{i+1} \le x \le 1, \end{cases}$$

and

$$l_{N+1}(x) \equiv \begin{cases} (x - x_N)(1 - x_N)^{-1}, & x_N \le x \le 1, \text{ and} \\ 0, & 0 \le x \le x_N. \end{cases}$$

In the special case of a uniform partition with mesh length $h = (N + 1)^{-1}$, the basis functions $l_i(x)$, $0 \le i \le N + 1$, can be expressed in terms of one "standard" basis function, $L(x)$. In fact, if

$$L(x) \equiv \begin{cases} 1 + x, & -1 \le x \le 0, \\ 1 - x, & 0 \le x \le 1, \\ 0, & x \in R - [-1, 1], \end{cases}$$

then

$$l_i(x) = L(h^{-1}x - i), \qquad 0 \le i \le N + 1.$$

Furthermore, the mapping $\vartheta_{L(\Delta)}$ is "local" in the sense that if $x \in [x_i, x_{i+1}]$, $0 \le i \le N$, then $\vartheta_{L(\Delta)}\mathbf{f}(x)$ depends only on f_i and f_{i+1}. If $f(x)$ is defined for all $x \in I$, we will let $\vartheta_{L(\Delta)} f \equiv \vartheta_{L(\Delta)}\mathbf{f}$, where $\mathbf{f} \equiv (f(x_0), \ldots, f(x_{N+1}))$. Moreover, we will often abbreviate $\vartheta_{L(\Delta)}$ by ϑ_L.

Any single evaluation of $\vartheta_L\mathbf{f}(x)$ requires only three multiplications and four additions. In fact, if $x \in [x_i, x_{i+1}]$, $0 \le i \le N$, then

$$\vartheta_L\mathbf{f}(x) = (x_{i+1} - x_i)^{-1}[f_i(x_{i+1} - x) + f_{i+1}(x - x_i)].$$

To get an idea of the behavior of the error in this procedure, we consider the simple function $f(x) \equiv x^2$. If $e_i(x) \equiv f(x) - \vartheta_L f(x)$, $x \in [x_i, x_{i+1}]$, $0 \le i \le N$, then

$$e_i(x) = x^2 - x_i^2(x_{i+1} - x)(x_{i+1} - x_i)^{-1} - x_{i+1}^2(x - x_i)(x_{i+1} - x_i)^{-1}.$$

Moreover, it is easily directly verified that

$$-\tfrac{1}{4}(x_{i+1} - x_i)^2 \le e_i(x) \le 0$$

for all $x \in [x_i, x_{i+1}]$, with the minimum being assumed at $x = \frac{1}{2}(x_i + x_{i+1})$. Thus, $\| f - \vartheta_L f \|_\infty \le \frac{1}{4}h^2$, which shows that for $f(x) = x^2$ the piecewise linear interpolation procedure is second-order accurate, i.e., the exponent of h in the error bound is 2. Using this bound we may compute the following table for uniform partitions $\Delta(h)$:

h	$\dim L(\Delta(h))$	$\|\|x^2 - \vartheta_{L(\Delta(h))}x^2\|\|_\infty$
1	2	0.25
10^{-1}	12	0.25×10^{-2}
10^{-2}	102	0.25×10^{-4}
10^{-3}	1002	0.25×10^{-6}

In Section 2.3, we will show that this special result generalizes and that this procedure is second-order accurate for all sufficiently smooth functions.

2.2 TWO-DIMENSIONAL PROBLEMS

In this section, we introduce a two-dimensional analogue of the interpolation procedure of the previous section.

We let $L(\rho) \equiv L(\Delta) \otimes L(\Delta_y)$ (the tensor product), i.e., $L(\rho)$ is the $(N+2)(M+2)$-dimensional linear space of all functions of the form

$$l(x, y) = \sum_{i=0}^{N+1} \sum_{j=0}^{M+1} c_{ij} l_i(x) l_j(y).$$

Clearly $L(\rho)$ can be characterized as the vector space of all continuous, piecewise bilinear polynomials with respect to ρ. In fact, if $\phi(x, y)$ is such a function, then for each $0 \leq i \leq N$, $0 \leq j \leq M$, and $(x, y) \in [x_i, x_{i+1}] \times [y_j, y_{j+1}]$,

$$\phi(x, y) = \phi(x_i, y_j) l_i(x) l_j(y) + \phi(x_{i+1}, y_j) l_{i+1}(x) l_j(y) + \phi(x_i, y_{j+1}) l_i(x) l_{j+1}(y) + \phi(x_{i+1}, y_{j+1}) l_{i+1}(x) l_{j+1}(y).$$

Thus,

$$\phi(x, y) = \sum_{i=0}^{N+1} \sum_{j=0}^{M+1} \phi(x_i, y_j) l_i(x) l_j(y),$$

and hence $\phi \in L(\rho)$.

Given the vector $\mathbf{f} \equiv \{f_{ij}\}_{i=0, j=0}^{N+1, M+1}$, we define

$$\vartheta_{L(\rho)} \mathbf{f} \equiv \sum_{i=0}^{N+1} \sum_{j=0}^{M+1} f_{ij} l_i(x) l_j(y) \tag{2.4}$$

as the interpolation mapping into $L(\rho)$. If $f_{ij} \equiv f(x_i, y_j)$, $0 \leq i \leq N+1$ and $0 \leq j \leq M+1$, where $f(x, y)$ is defined for all $(x, y) \in R$, we will write

$\vartheta_{L(\rho)} f$ for $\vartheta_{L(\rho)} \mathbf{f}$. We now give a characterization of $\vartheta_{L(\rho)} f$ in terms of one-dimensional interpolation schemes.

THEOREM 2.1

If $f(x, y)$ is defined for all $(x, y) \in U$, then

$$\vartheta_{L(\rho)} f = \vartheta_{L(\Delta_y)} \vartheta_{L(\Delta)} f = \vartheta_{L(\Delta)} \vartheta_{L(\Delta_y)} f. \tag{2.5}$$

Proof. We prove only the first equality in (2.5), as the second is proved the same way. By definition

$$\begin{aligned}
\vartheta_{L(\Delta_y)} \vartheta_{L(\Delta)} f &= \vartheta_{L(\Delta_y)} \left[\sum_{i=0}^{N+1} f(x_i, y) l_i(x) \right] \\
&= \sum_{j=0}^{M+1} \left(\sum_{i=0}^{N+1} f(x_i, y_j) l_i(x) \right) l_j(y) \\
&= \sum_{i=0}^{N+1} \sum_{j=0}^{M+1} f(x_i, y_j) l_i(x) l_j(y) \\
&= \vartheta_{L(\rho)} f.
\end{aligned}$$

Q.E.D.

2.3 ERROR ANALYSIS

In this section, we prove a priori error bounds for the interpolation procedures introduced in Sections 2.1 and 2.2. In the one-dimensional case, our analysis is based upon the fact that the piecewise linear interpolating function describes the shape of a taut string passing through the interpolation points. As such, it can be characterized as the solution of a simple variational problem.

This is also true of the piecewise cubic interpolation procedures of Chapters 3 and 4. Hence, we treat the piecewise linear interpolation procedure in full detail as a guide to the higher-order cases.

We now state and prove a variational characterization of the piecewise linear interpolate $\vartheta_{L(\Delta)} \mathbf{f}$ as the interpolating function of minimum least squares variation.

THEOREM 2.2

Let Δ and $\{f_i\}_{i=0}^{N+1}$ be given and

$$V \equiv \{w \in PC^{1,2}(I) \mid w(x_i) = f_i,\ 0 \le i \le N+1\}.$$

The variational problem of finding the functions $p \in V$ which minimize $\| Dw \|_2^2$ over all $w \in V$ has the unique solution $\vartheta_{L(\Delta)} \mathbf{f}$.

Proof. First, we show that $p \in V$ is a solution of the variational problem if and only if

$$(2.6) \qquad (Dp, D\delta)_2 = 0$$

for all $\delta \in V_0 \equiv \{w \in PC^{1,2}(I) \mid w(x_i) = 0, 0 \leq i \leq N + 1\}$, i.e., if and only if p is a solution of the generalized Euler equation.

In fact, if $p \in V$, then $p + \alpha\delta \in V$ for all real numbers α and all $\delta \in V_0$. Moreover, if p is a solution of the variational problem, the function

$$(2.7) \quad F(\alpha) \equiv \|D(p + \alpha\delta)\|_2^2 = (Dp, Dp)_2 + 2\alpha(Dp, D\delta)_2 + \alpha^2(D\delta, D\delta)_2$$

is minimized for $\alpha = 0$. Thus, by calculus, $dF(0)/d\alpha = 0$ and we obtain (2.6). Conversely, if $p \in V$ is a solution of (2.6) and $w \in V$, then $w - p \in V_0$ and $(Dp, Dw - Dp)_2 = 0$. Thus, by the corollary to Theorem 1.4, we have

$$(2.8) \qquad \|D(w - p)\|_2^2 + \|Dp\|_2^2 = \|Dw\|_2^2.$$

Hence,

$$(2.9) \qquad \|Dp\|_2^2 \leq \|Dw\|_2^2 \qquad \text{for all } w \in V$$

and p is a solution of the variational problem. Moreover, we have equality in (2.9) if and only if $\|Dw - Dp\|_2^2 = 0$ or, using the Rayleigh–Ritz Inequality (Theorem 1.2), if and only if

$$\pi^2 \|w - p\|_2^2 \leq \|D(w - p)\|_2^2 = 0,$$

or $w = p$. Thus, the variational problem and the generalized Euler equation (2.6) have at most one solution.

Second, we complete our proof by showing that $\vartheta_{L(\Delta)}\mathbf{f}$ is a solution of the generalized Euler equation (2.6). If $\delta \in V_0$, then integrating by parts we have

$$\begin{aligned}(D\vartheta_{L(\Delta)}\mathbf{f}, D\delta)_2 &= \int_0^1 D\vartheta_{L(\Delta)}\mathbf{f}(x)D\delta(x)\,dx \\ &= \sum_{i=0}^{N} \int_{x_i}^{x_{i+1}} D\vartheta_{L(\Delta)}\mathbf{f}(x)D\delta(x)\,dx \\ &= \sum_{i=0}^{N} [\delta(x)D\vartheta_{L(\Delta)}\mathbf{f}(x)]_{x_i}^{x_{i+1}} - \sum_{i=0}^{N} \int_{x_i}^{x_{i+1}} \delta(x)D^2\vartheta_{L(\Delta)}\mathbf{f}(x)\,dx.\end{aligned}$$

Thus, $(D\vartheta_{L(\Delta)}\mathbf{f}, D\delta)_2 = 0$, since each term in the first sum vanishes because of the interpolation conditions on p, and each term of the second sum vanishes because $\vartheta_{L(\Delta)}\mathbf{f}$ is a linear polynomial on each subinterval $[x_i, x_{i+1}]$, $0 \leq i \leq N$. Q.E.D.

As a corollary of equation (2.8) of the preceding proof we obtain the so-called "First Integral Relation," which was introduced in [2.1].

COROLLARY

If $f \in PC^{1,2}(I)$, then

$$\|D\vartheta_L f\|_2^2 + \|D\vartheta_L f - Df\|_2^2 = \|Df\|_2^2. \tag{2.10}$$

By using the same type of integration by parts argument that we used in the proof of Theorem 2.2, we may prove the following result.

THEOREM 2.3

If $g \in PC^{2,2}(I)$, and $g(x_i) = f_i$, $0 \leq i \leq N + 1$, then

$$\|D(g - \vartheta_L \mathbf{f})\|_2^2 = (g - \vartheta_L \mathbf{f}, D^2 g)_2. \tag{2.11}$$

For the special case in which $f_i = f(x_i)$, $0 \leq i \leq N + 1$, and $g = f$, we obtain the so-called "Second Integral Relation," which was introduced in [2.1].

COROLLARY

If $f \in PC^{2,2}(I)$, then

$$\|D(f - \vartheta_L f)\|_2^2 = (f - \vartheta_L f, D^2 f)_2. \tag{2.12}$$

We now turn to the derivation of a priori bounds for the interpolation error, $f - \vartheta_L f$, and its derivative with respect to the L^2-norm and the L^∞-norm. In general, variational problems lead naturally to error bounds in the L^2-norm. However, for computational purposes we prefer L^∞-norm error bounds. We start with a preliminary error bound which is not only of interest for its own sake, but will be used in the remainder of this chapter.

THEOREM 2.4

If $f \in PC^{1,2}(I)$, then

$$\|D(f - \vartheta_L f)\|_2 \leq \|Df\|_2 \tag{2.13}$$

and

$$\|f - \vartheta_L f\|_2 \leq \pi^{-1} h \|Df\|_2. \tag{2.14}$$

Proof. Inequality (2.13) follows directly from the First Integral Relation (2.10). To prove inequality (2.14) we note that $f(x_i) - \vartheta_L f(x_i) = 0$, for all $0 \leq i \leq N + 1$, and by the Rayleigh–Ritz Inequality (Theorem 1.2)

$$(2.15)\quad \int_{x_i}^{x_{i+1}} [f(x) - \vartheta_L f(x)]^2\, dx \le \pi^{-2}(x_{i+1} - x_i)^2 \int_{x_i}^{x_{i+1}} [Df(x) - D\vartheta_L f(x)]^2\, dx,$$

for all $0 \le i \le N$.

Summing both sides of inequality (2.15) with respect to i from 0 to N and taking the square root of both sides of the resulting inequality, we obtain

$$(2.16)\qquad \| f - \vartheta_L f \|_2 \le \pi^{-1} h \| D(f - \vartheta_L f) \|_2,$$

and (2.14) follows by using (2.13) to bound the right hand side of (2.16).
Q.E.D.

If f is somewhat smoother, we may obtain stronger a priori bounds. The "boot-strap" method of proof, which we will give, was suggested in [2.1] and refined in [2.5].

THEOREM 2.5

If $f \in PC^{2,2}(I)$, then

$$(2.17)\qquad \| D(f - \vartheta_L f) \|_2 \le \pi^{-1} h \| D^2 f \|_2$$

and

$$(2.18)\qquad \| f - \vartheta_L f \|_2 \le \pi^{-2} h^2 \| D^2 f \|_2.$$

Proof. Applying the Cauchy–Schwarz Inequality to the second integral relation (2.12) yields

$$(2.19)\qquad \| D(f - \vartheta_L f) \|_2^2 \le \| D^2 f \|_2 \| f - \vartheta_L f \|_2.$$

Combining this with (2.16), we obtain (2.17).

Now using (2.17) to bound the right-hand side of (2.16), we obtain (2.18).
Q.E.D.

We now turn to a derivation of error bounds in the L^∞-norm.

THEOREM 2.6

If $f \in PC^{2,\infty}(I)$, then

$$(2.20)\qquad \| f - \vartheta_L f \|_\infty \le \tfrac{1}{8} h^2 \| D^2 f \|_\infty$$

and

$$(2.21)\qquad \| D(f - \vartheta_L f) \|_\infty \le \tfrac{1}{2} h \| D^2 f \|_\infty.$$

Proof. For fixed $0 \leq i \leq N$, let $w(x) \equiv (x - x_i)(x - x_{i+1})$. For each $x \in [x_i, x_{i+1}]$, there exists $\xi_x \in [x_i, x_{i+1}]$ such that

$$e(x) \equiv f(x) - \vartheta_L f(x) = \tfrac{1}{2} D^2 f(\xi_x) w(x).$$

In fact, if $x = x_i$ or x_{i+1}, any point ξ_x suffices. Otherwise, for fixed $\tilde{x}$, choose λ such that

$$\theta(\tilde{x}) \equiv e(\tilde{x}) - \lambda w(\tilde{x}) = 0.$$

But $\theta(x)$ has three zeroes in $[x_i, x_{i+1}]$ and hence by Rolle's Theorem there exists a point $\xi_x \in [x_i, x_{i+1}]$ such that $D^2\theta(\xi_x) = 0$. But $D^2\theta(\xi_x) = D^2 f(\xi_x) - 2\lambda$ and hence $\lambda = \frac{1}{2} D^2 f(\xi_x)$. Thus,

$$\max_{x \in [x_i, x_{i+1}]} |e(x)| \leq \tfrac{1}{2} \|D^2 f\|_\infty \max_{x \in [x_i, x_{i+1}]} |w(x)| \leq \tfrac{1}{8} h^2 \|D^2 f\|_\infty,$$

which proves (2.20).

We now give an alternate proof of (2.20) based on the Peano Kernel Theorem. Applying this theorem to the functional $e(x)$ for fixed $x \in [x_i, x_{i+1}]$, we have

$$e(x) = \int_{x_i}^{x_{i+1}} K_x(t) D^2 f(t)\, dt,$$

where

$$K_x(t) \equiv \begin{cases} (x_{i+1} - x)(t - x_i)(x_{i+1} - x_i)^{-1}, & x_i \leq t \leq x \leq x_{i+1}, \\ (x - x_i)(x_{i+1} - t)(x_{i+1} - x_i)^{-1}, & x_i \leq x \leq t \leq x_{i+1}. \end{cases}$$

Thus,

$$\begin{aligned} |e(x)| &\leq \|D^2 f\|_\infty (x_{i+1} - x_i)^{-1} \Big[\int_{x_i}^{x} (x_{i+1} - x)(t - x_i)\, dt \\ &\qquad + \int_{x}^{x_{i+1}} (x - x_i)(x_{i+1} - t)\, dt \Big] \\ &= \tfrac{1}{2} \|D^2 f\|_\infty (x_{i+1} - x_i)^{-1} [(t - x_i)^2 (x_{i+1} - x)|_{x_i}^{x} \\ &\qquad - (t - x_{i+1})^2 (x - x_i)|_{x}^{x_{i+1}}] \\ &= \tfrac{1}{2} \|D^2 f\|_\infty (x_{i+1} - x_i)^{-1} [(x - x_i)^2 (x_{i+1} - x) \\ &\qquad + (x - x_{i+1})^2 (x - x_i)]. \end{aligned}$$

By a standard application of the differential calculus, we find that the maximum value of the expression in brackets occurs at $x = (x_i + x_{i+1})/2$ and is equal to $\frac{1}{4}(x_{i+1} - x_i)^2$, which proves (2.20).

Similarly,

$$\begin{aligned} D_x e(x) = &\int_{x_i}^{x_{i+1}} D_x K_x(t) D^2 f(t)\, dt \\ &+ (x_{i+1} - x_i)^{-1} [(x_{i+1} - x)(x - x_i) - (x - x_i)(x_{i+1} - x)] \end{aligned}$$

$$\leq (x_{i+1} - x_i)^{-1} \|D^2 f\|_\infty \left[\int_{x_i}^{x} (t - x)\, dt + \int_{x}^{x_{i+1}} (x_{i+1} - t)\, dt \right]$$
$$= \tfrac{1}{2}(x_{i+1} - x_i)^{-1} \|D^2 f\|_\infty [(t - x_i)^2 |_{x_i}^{x} - (x_{i+1} - t)^2 |_{x}^{x_{i+1}}]$$
$$= \tfrac{1}{2}(x_{i+1} - x_i)^{-1} \|D^2 f\|_\infty [(x - x_i)^2 + (x_{i+1} - x)^2].$$

The maximum value of the expression in brackets occurs at either $x = x_i$ or $x = x_{i+1}$ and is equal to $(x_{i+1} - x_i)^2$, which proves (2.21). Q.E.D.

The preceding proof shows that the interpolation error $f(x) - \vartheta_L f(x)$ for $x \in [x_i, x_{i+1}]$ depends only on values of $D^2 f(t)$ for $t \in [x_i, x_{i+1}]$. Furthermore, if f is sufficiently smooth, then the interpolation mapping, ϑ_L, produces an approximation which is second-order accurate with respect to both the L^∞-norm and the L^2-norm, i.e., the exponent of h in the error bounds (2.18) and (2.20) is 2.

We now proceed to the a priori error bounds for the piecewise bilinear interpolation procedure. As in the one-dimensional case, we find that if f is sufficiently smooth, then the interpolate $\vartheta_{L(\rho)} f$ is a second-order approximation to f with respect to both the L^∞-norm and the L^2-norm.

THEOREM 2.7

If $f \in PC^{2,2}(U)$, then

$$\begin{aligned} \|f - \vartheta_{L(\rho)} f\|_2 &\leq \pi^{-2}(h^2 \|D_x^2 f\|_2 + hk \|D_x D_y f\|_2 + k^2 \|D_y^2 f\|_2) \\ &\leq \pi^{-2} \bar{\rho}^2(\|D_x^2 f\|_2 + \|D_x D_y f\|_2 + \|D_y^2 f\|_2), \end{aligned} \tag{2.22}$$

$$\begin{aligned} \|D_x(f - \vartheta_{L(\rho)} f)\|_2 &\leq \pi^{-1}(h \|D_x^2 f\|_2 + 2k \|D_x D_y f\|_2) \\ &\leq \pi^{-1} \bar{\rho}(\|D_x^2 f\|_2 + 2 \|D_x D_y f\|_2), \end{aligned} \tag{2.23}$$

and

$$\begin{aligned} \|D_y(f - \vartheta_{L(\rho)} f)\|_2 &\leq \pi^{-1}(k \|D_y^2 f\|_2 + 2h \|D_x D_y f\|_2) \\ &\leq \pi^{-1} \bar{\rho}(\|D_y^2 f\|_2 + 2 \|D_x D_y f\|_2). \end{aligned} \tag{2.24}$$

Proof. From (2.5) and the triangle inequality, we have

$$\begin{aligned} \|f - \vartheta_{L(\rho)} f\|_2 &\leq \|f - \vartheta_{L(\Delta)} f\|_2 + \|\vartheta_{L(\Delta)}(f - \vartheta_{L(\Delta_y)} f)\|_2 \\ &\leq \|f - \vartheta_{L(\Delta)} f\|_2 + \|\vartheta_{L(\Delta)}(f - \vartheta_{L(\Delta_y)} f) \\ &\quad - (f - \vartheta_{L(\Delta_y)} f)\|_2 + \|f - \vartheta_{L(\Delta_y)} f\|_2. \end{aligned} \tag{2.25}$$

Using the results of Theorems 2.4 and 2.5 to bound the right-hand side of (2.25), we have

$$\begin{aligned} \|f - \vartheta_{L(\rho)} f\|_2 &\leq \pi^{-2} h^2 \|D_x^2 f\|_2 \\ &\quad + \pi^{-1} h \|D_x(f - \vartheta_{L(\Delta_y)} f)\|_2 + \pi^{-2} k^2 \|D_y^2 f\|_2. \end{aligned} \tag{2.26}$$

But since $D_x \vartheta_{L(\Delta_y)} f = \vartheta_{L(\Delta_y)} D_x f$, we have

$$\|D_x(f - \vartheta_{L(\Delta_y)} f)\|_2 \leq \pi^{-1} k \|D_y D_x f\|_2. \tag{2.27}$$

Using (2.27) to bound the right-hand side of (2.26), we obtain (2.22).

We may prove (2.23) and (2.24) in a similar way. For example, using the results of Theorems 2.4 and 2.5 we have

$$\begin{aligned}\|D_x(f - \vartheta_{L(\rho)} f)\|_2 &\leq \|D_x(f - \vartheta_{L(\Delta)} f)\|_2 + \|D_x[\vartheta_{L(\Delta)}(f - \vartheta_{L(\Delta_y)} f) \\ &\qquad - (f - \vartheta_{L(\Delta_y)} f)]\|_2 + \|D_x(f - \vartheta_{L(\Delta_y)} f)\|_2 \\ &\leq \pi^{-1} h \|D_x^2 f\|_2 + 2\|D_x(f - \vartheta_{L(\Delta_y)} f)\|_2 \\ &\leq \pi^{-1} h \|D_x^2 f\|_2 + 2\pi^{-1} k \|D_x D_y f\|_2,\end{aligned}$$

which proves (2.23). Q.E.D.

The error bounds for the L^∞-norm are proved in a similar way using the following result, which is of interest for its own sake.

LEMMA 2.1

$$\|\vartheta_{L(\Delta)} \mathbf{f}\|_\infty \leq \max_{0 \leq i \leq N+1} |f_i| \equiv \|\mathbf{f}\|_\infty.$$

Proof. If $x \in [x_j, x_{j+1}]$, $0 \leq j \leq N$,

$$\begin{aligned}|\vartheta_{L(\Delta)} \mathbf{f}(x)| &= |f_j(x_{j+1} - x)(x_{j+1} - x_j)^{-1} + f_{j+1}(x - x_j)(x_{j+1} - x_j)^{-1}| \\ &\leq \|\mathbf{f}\|_\infty [(x_{j+1} - x) + (x - x_j)](x_{j+1} - x_j)^{-1} \\ &= \|\mathbf{f}\|_\infty.\end{aligned}$$

Q.E.D.

The result of Lemma 2.1 shows that $\vartheta_{L(\Delta)} \mathbf{f}$ depends continuously on $\mathbf{f}$ in a very strong way. Moreover, if $\mathbf{f}^*$ is an approximation to $\mathbf{f}$ such that $\|\mathbf{f} - \mathbf{f}^*\|_\infty \leq \epsilon$, then

$$\|\vartheta_{L(\Delta)} \mathbf{f} - \vartheta_{L(\Delta)} \mathbf{f}^*\|_\infty \leq \epsilon.$$

THEOREM 2.8

If $f \in PC^{2,\infty}(U)$, then

$$\begin{aligned}\|f - \vartheta_{L(\rho)} f\|_\infty &\leq \tfrac{1}{8}(h^2 \|D_x^2 f\|_\infty + k^2 \|D_y^2 f\|_\infty) \\ &\leq \tfrac{1}{8}\bar{\rho}^2(\|D_x^2 f\|_\infty + \|D_y^2 f\|_\infty).\end{aligned} \tag{2.28}$$

Proof. Using (2.5) and Lemma 2.1, we have

$$\begin{aligned}\|f - \vartheta_{L(\rho)} f\|_\infty &\leq \|f - \vartheta_{L(\Delta)} f\|_\infty + \|\vartheta_{L(\Delta)}(f - \vartheta_{L(\Delta_y)} f)\|_\infty \\ &\leq \|f - \vartheta_{L(\Delta)} f\|_\infty + \|f - \vartheta_{L(\Delta_y)} f\|_\infty,\end{aligned}$$

and (2.28) follows by using the result of Theorem 2.6 to bound the right-hand side of this inequality. Q.E.D.

EXERCISES FOR CHAPTER 2

(2.1) Let $f \in PC^{1,\infty}(I)$ and $\tilde{f}(x) \equiv \{f[(x_i + x_{i+1})/2] \mid x \in [x_i, x_{i+1}], 0 \leq i \leq N\}$ be the piecewise constant interpolate of $f(x)$. Using the Mean Value Theorem of calculus, show that $\|f - \tilde{f}\|_\infty \leq \frac{1}{2}h \|Df\|_\infty$.

(2.2) Let $f \in PC^{1,\infty}(U)$ and $\tilde{f}(x, y) \equiv \{f[(x_i + x_{i+1})/2, (y_j + y_{j+1})/2] \mid (x, y) \in [x_i, x_{i+1}] \times [y_j, y_{j+1}], 0 \leq i \leq N, 0 \leq j \leq M\}$ be the piecewise constant interpolate of $f(x, y)$. Show that

$$\|f - \tilde{f}\|_\infty \leq \tfrac{1}{2}\{h \|D_x f\|_\infty + k \|D_y f\|_\infty\}.$$

(2.3) Use the Peano Kernel Theorem to show that

$$\|f - \vartheta_{L(\Delta)} f\|_\infty \leq \tfrac{1}{2}h \|Df\|_\infty \qquad \text{for all } f \in PC^{1,\infty}(I).$$

(2.4) Use the result of Exercise (2.3) to show that

$$\|f - \vartheta_{L(\rho)} f\|_\infty \leq \tfrac{1}{2}\{h \|D_x f\|_\infty + k \|D_y f\|_\infty\} \qquad \text{for all } f \in PC^{1,\infty}(U).$$

(2.5) Let $\Delta(h): 0 < h < 2h < \ldots < (N+1)h = 1$ be the uniform mesh of mesh length h. Show that $\vartheta_{L(\Delta(h))} \sin h^{-1}\pi x = 0$ and explicitly evaluate both sides of the error bounds (2.14) and (2.15). What can you conclude about the exponent of h and the constant factor in (2.14) and (2.15) from this example?

(2.6) Let $\theta(x) \equiv \{(-1)^i(x - x_i)(x_{i+1} - x)(x_{i+1} - x_i)^{-1} \mid x \in [x_i, x_{i+1}], 0 \leq i \leq N\}$. Show that $\theta(x) \in PC^{2,2}(I)$, $\vartheta_{L(\Delta)}\theta(x) \equiv 0$, and explicitly evaluate both sides of the error bounds (2.17) and (2.18). What can you conclude about the exponent of h and the constant factor in (2.17) and (2.18)?

(2.7) Let

$$\Delta_{\alpha,N}: 0 < (N+1)^{-q} < \ldots < j^q (N+1)^{-q} < \ldots < (N+1)^q(N+1)^{-q} = 1,$$

where $q \equiv 2\alpha^{-1}$, be a partition of [0, 1] for all $0 < \alpha < 2$, $\alpha \neq 1$. Show that if $E_j \equiv \max_{x \in [x_j, x_{j+1}]} |x^\alpha - \vartheta_{L(\Delta_{\alpha,N})} x^\alpha|$, $0 \leq j \leq N$, then $E_0 \leq (N+1)^{-2}$ and

$$E_j \leq \frac{q^2}{8}\left(\frac{j+1}{j}\right)^{2(q-1)} (N+1)^{-2}\alpha|\alpha - 1|, \qquad 1 \leq j \leq N.$$

Thus, if we choose the partition properly, we can define an interpolation scheme for x^α, $0 < \alpha < 2$, $\alpha \neq 1$, which is second-order accurate in the L^∞-norm, even though $x \in PC^{2,\infty}(I)$.

$$\|x^\alpha - \vartheta_{L(\Delta_{\alpha,N})} x^\alpha\|_\infty \leq \max\left(1, \frac{q^2}{16}4^q\alpha|\alpha - 1|\right)(N+1)^{-2}$$

(cf. [2.4]).

(2.8) Use the Peano Kernel Theorem to show that if $f \in PC^{t,q}(I)$, $t = 1$ or 2, then there exists a positive constant, K, which can be explicitly computed, such that

$$\|f - \vartheta_L f\|_p \leq \begin{cases} Kh^{t+p^{-1}-q^{-1}} \|D^t f\|_q, & \text{if } p \geq q \geq 1, \\ Kh^t \|D^t f\|_q, & \text{if } q \geq p \geq 1, \end{cases}$$

for all partitions Δ of I.

(2.9) Show that if $D^2 f(x) \geq 0$ for all $x \in [x_k, x_{k+1}]$ for some $0 \leq k \leq N$, then $\vartheta_L f(x) \geq f(x)$ for all $x \in [x_k, x_{k+1}]$.

(2.10) Use Exercises (1.2) and (1.4) to show that if $f \in PC^{1,2}(I)$, then for all $p \geq 2$,

$$\|f - \vartheta_L f\|_p \leq \tfrac{1}{2} h^{1/2+p^{-1}} \|Df\|_2,$$

and if $f \in PC^{2,2}(I)$, then for all $p \geq 2$,

$$\|f - \vartheta_L f\|_p \leq (2\pi)^{-1} h^{3/2+p^{-1}} \|D^2 f\|_2.$$

(2.11) Show that if $f \in PC^{1,2}(I)$, then for all $p \leq 2$,

$$\|f - \vartheta_L f\|_p \leq \pi^{-1} h \|Df\|_2,$$

and if $f \in PC^{2,2}(I)$, then for all $p \leq 2$,

$$\|f - \vartheta_L f\|_p \leq \pi^{-2} h^2 \|Df\|_2.$$

(2.12) Show that if $f \in PC^{1,2}(I)$, then $\vartheta_L f$ satisfies the "local First Integral Relation,"

$$\int_{x_i}^{x_{i+1}} (D\vartheta_L f(x))^2\,dx + \int_{x_i}^{x_{i+1}} (D\vartheta_L f(x) - Df(x))^2\,dx = \int_{x_i}^{x_{i+1}} (Df(x))^2\,dx, \qquad 0 \leq i \leq N.$$

(2.13) Show that if $f \in PC^{2,2}(I)$, then $\vartheta_L f$ satisfies the "local Second Integral Relation,"

$$\int_{x_i}^{x_{i+1}} (Df(x) - D\vartheta_L f(x))^2\,dx = \int_{x_i}^{x_{i+1}} (f(x) - \vartheta_L f(x)) D^2 f(x)\,dx \qquad 0 \leq i \leq N.$$

(2.14) Using the results of Exercises (2.12) and (2.13), prove local versions of the results of Theorems 2.3 and 2.4.

(2.15) Consider the trapezoidal rule quadrature scheme

$$\int_0^1 f(x)\,dx \approx \int_0^1 \vartheta_{L(\Delta)} f(x)\,dx.$$

Show that for a uniform mesh $\Delta(h)$ this scheme reduces to

$$\int_0^1 f(x)\,dx \approx \frac{h}{2}(f(0) + 2\sum_{i=1}^{N} f(x_i) + f(1)).$$

Develop an analogous formula for the case of nonuniform meshes.

REFERENCES FOR CHAPTER 2

[2.1] AHLBERG, J. H., E. N. NILSON, and J. L. WALSH, Convergence properties of generalized splines. *Proc. Nat. Acad. Sci. U.S.A.* **54**, 344–350 (1965).

[2.2] BRAMBLE, J., and S. HILBERT, Bounds for a class of linear functionals with application to Hermite Interpolation. *Numer. Math.* **16**, 362–369 (1971).

[2.3] CHENEY, E. W., *Introduction to Approximation Theory*, McGraw-Hill, New York (1966).

[2.4] RICE, John R., On the degree of convergence of nonlinear spline approximation. *Approximations with Special Emphasis on Spline Functions* (I. J. Schoenberg, ed.) 349–367, Academic Press, New York (1969).

[2.5] SCHULTZ, M. H., and R. S. VARGA, *L*-splines. *Numer. Math.* **10**, 345–369 (1967).

3 PIECEWISE CUBIC HERMITE INTERPOLATION

3.1 ONE-DIMENSIONAL PROBLEMS

In this chapter, we introduce and study a Hermite interpolation procedure which is fourth-order accurate. Given Δ: $0 = x_0 < x_1 < \ldots < x_N < x_{N+1} = 1$ and $2(N+2)$ real numbers, $\{f_i, f_i^1\}_{i=0}^{N+1}$, a Hermite interpolating procedure yields a function, $g(x)$, such that $g(x)$ is defined for all $x \in I$, $g(x_i) = f_i$, $0 \leq i \leq N+1$, and $Dg(x_i) = f_i^1$, $0 \leq i \leq N+1$.

In order to have a "good" procedure, we extend the ideas of Chapter 2 and consider a piecewise cubic Hermite interpolation procedure. We begin with two basic definitions.

DEFINITION 3.1

Given Δ, let $H(\Delta)$ be the $2(N+2)$-dimensional vector space of all continuously differentiable, piecewise cubic polynomials with respect to Δ, i.e.,

$$H(\Delta) \equiv \{p(x) \in C^1(I) \,|\, p(x) \text{ is a cubic polynomial on each subinterval } [x_i, x_{i+1}],\ 0 \leq i \leq N, \text{ defined by } \Delta\}.$$

DEFINITION 3.2

Given $\mathbf{f} \equiv (f_0, f_0^1, f_1, f_1^1, \ldots, f_{N+1}, f_{N+1}^1) \in R^{2N+4}$, let $\vartheta_{H(\Delta)}\mathbf{f}$, the $H(\Delta)$-*interpolate of* $\mathbf{f}$, be the unique element, $h(x)$, in $H(\Delta)$ such that $h(x_i) = f_i$, $0 \leq i \leq N+1$, and $Dh(x_i) = f_i^1$, $0 \leq i \leq N+1$.

This procedure is well-defined. In fact, if $h(x) \in H(\Delta)$ interpolates $\mathbf{f}$ as above, then $e(x) \equiv \vartheta_{H(\Delta)}\mathbf{f}(x) - h(x)$ is a cubic polynomial on $[x_i, x_{i+1}]$, $0 \leq i \leq N$, and $e(x_i) = De(x_i) = e(x_{i+1}) = De(x_{i+1}) = 0$, $0 \leq i \leq N$, which implies that $e(x) = c(x - x_i)^2(x - x_{i+1})^2$ for some constant c and all $x \in [x_i, x_{i+1}]$. Since $e(x)$ is a cubic polynomial, c must be zero and $e(x) \equiv 0$ for all $x \in I$. We can express the $H(\Delta)$-interpolate as

$$\vartheta_{H(\Delta)}\mathbf{f} = \sum_{i=0}^{N+1} (f_i h_i(x) + f_i^1 h_i^1(x)),$$

where $h_i(x)$ is the unique element in $H(\Delta)$ such that $h_i(x_j) = \delta_{ij}$, $0 \leq i, j \leq N + 1$, and $Dh_i(x_j) = 0$, $0 \leq i, j \leq N + 1$, and $h_i^1(x)$ is the unique element in $H(\Delta)$ such that $h_i^1(x_j) = 0$, $0 \leq i, j \leq N + 1$, and $Dh_i^1(x_j) = \delta_{ij}$, $0 \leq i, j \leq N + 1$. The graph of $h_0(x)$ is given by

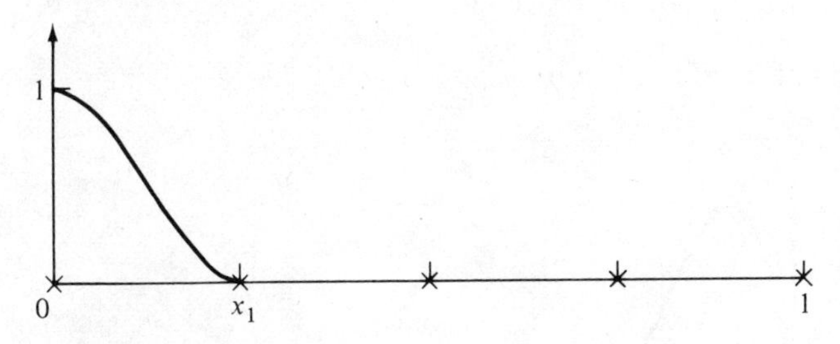

that of $h_i(x)$, $1 \leq i \leq N$, is given by

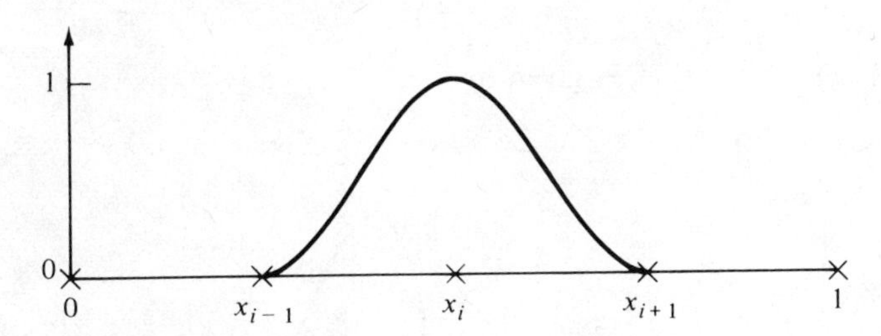

and that of $h_{N+1}(x)$ is given by

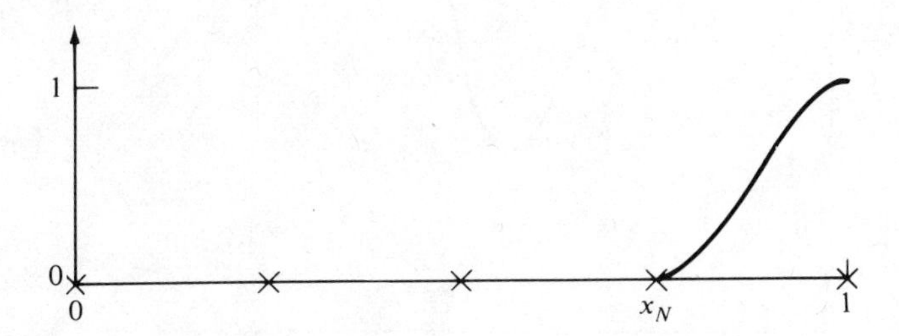

Moreover,

$$h_0(x) \equiv \begin{cases} 2x_1^{-3}x^3 - 3x_1^{-2}x^2 + 1, & 0 \leq x \leq x_1, \\ 0, & x_1 \leq x \leq 1, \end{cases}$$

$$h_i(x) \equiv \begin{cases} -2(x_i - x_{i-1})^{-3}(x - x_{i-1})^3 + 3(x_i - x_{i-1})^{-2}(x - x_{i-1})^2, \\ \hfill x_{i-1} \le x \le x_i, \\ 2(x_{i+1} - x_i)^{-3}(x - x_i)^3 - 3(x_{i+1} - x_i)^{-2}(x - x_i)^2 + 1, \\ \hfill x_i \le x \le x_{i+1}, \\ 0, \hfill x \in [0, 1] - [x_{i-1}, x_{i+1}], \end{cases}$$

for $1 \le i \le N$, and

$$h_{N+1}(x) \equiv \begin{cases} -2(1 - x_N)^{-3}(x - x_N)^3 + 3(1 - x_N)^{-2}(x - x_N)^2, & x_N \le x \le 1, \\ 0, & 0 \le x \le x_N. \end{cases}$$

The graph of $h_0^1(x)$ is given by

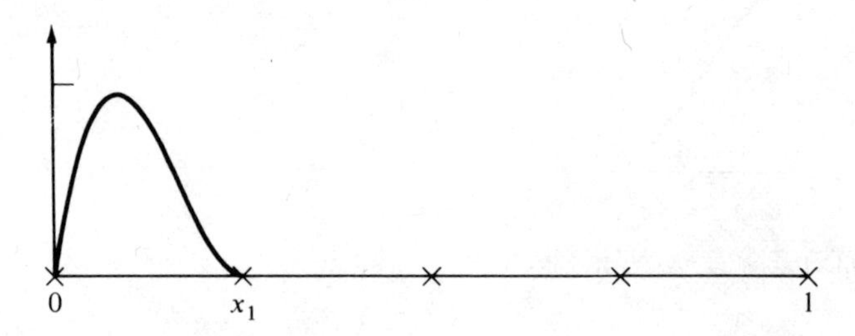

that of $h_i^1(x)$, $1 \le i \le N$, is given by

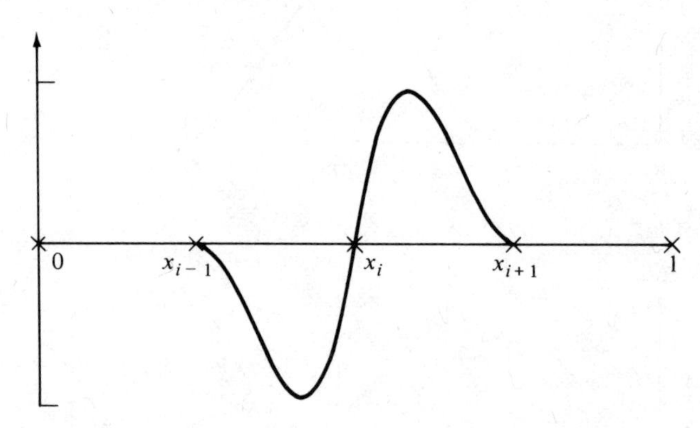

and that of $h_{N+1}^1(x)$ is given by

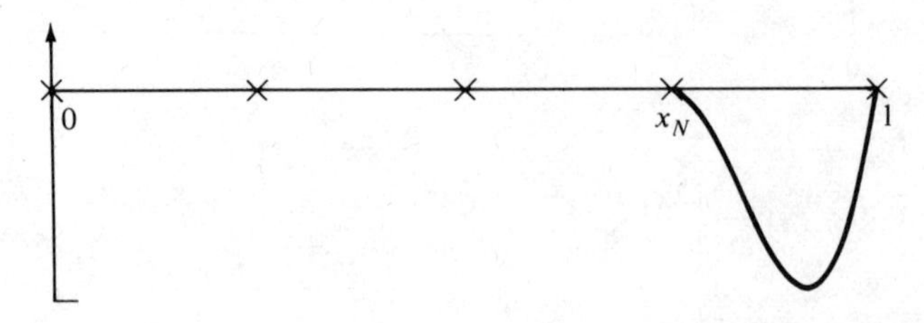

Moreover,

$$h_0^1(x) \equiv \begin{cases} x_1^{-2}x(x_1 - x)^2, & 0 \leq x \leq x_1, \\ 0, & x_1 \leq x \leq 1, \end{cases}$$

$$h_i^1(x) \equiv \begin{cases} (x_i - x_{i-1})^{-2}(x - x_{i-1})^2(x - x_i), & x_{i-1} \leq x \leq x_i, \\ (x_{i+1} - x_i)^{-2}(x - x_i)(x_{i+1} - x)^2, & x_i \leq x \leq x_{i+1}, \\ 0, & x \in [0, 1] - [x_{i-1}, x_{i+1}], \end{cases}$$

for $1 \leq i \leq N$, and

$$h_{N+1}^1(x) \equiv \begin{cases} (1 - x_N)^{-2}(x - x_N)^2(x - 1), & x_N \leq x \leq 1, \\ 0, & 0 \leq x \leq x_N. \end{cases}$$

Clearly, $\vartheta_{H(\Delta)}\mathbf{f}(x_i) = f_i$, $0 \leq i \leq N + 1$, and $D\vartheta_{H(\Delta)}\mathbf{f}(x_i) = f_i^1$, $0 \leq i \leq N + 1$.

In the special case of a uniform partition with mesh length $h = (N + 1)^{-1}$, the basis functions $h_i(x)$, $h_i^1(x)$, $0 \leq i \leq N + 1$ can be expressed in terms of two "standard" basis functions, $H(x)$ and $H^1(x)$. In fact, if

$$H(x) \equiv \begin{cases} (x + 1)^2(1 - 2x), & -1 \leq x \leq 0 \\ 2x^3 - 3x^2 + 1, & 0 \leq x \leq 1 \\ 0, & x \in R - [-1, 1], \end{cases}$$

and

$$H^1(x) \equiv \begin{cases} x(x + 1)^2, & -1 \leq x \leq 0 \\ x(1 - x)^2, & 0 \leq x \leq 1 \\ 0, & x \in R - [-1, 1], \end{cases}$$

then

$$h_i(x) = H(h^{-1}x - i), \; 0 \leq i \leq N + 1,$$

and

$$h_i^1(x) = hH^1(h^{-1}x - i), \; 0 \leq i \leq N + 1.$$

Furthermore, the mapping $\vartheta_{H(\Delta)}$ is "local" in the sense that if $x \in [x_i \, x_{i+1}]$, $0 \leq i \leq N$, then $\vartheta_{H(\Delta)}\mathbf{f}(x)$ depends only on f_i, f_i^1, f_{i+1}, and f_{i+1}^1.

If $f(x)$ and $Df(x)$ are defined for all $x \in I$, we will let $\vartheta_{H(\Delta)}f \equiv \vartheta_{H(\Delta)}\mathbf{f}$, where

$$\mathbf{f} \equiv (f(x_0), Df(x_0), f(x_1), \ldots, f(x_{N+1}), Df(x_{N+1})).$$

Moreover, we will usually abbreviate $\vartheta_{H(\Delta)}$ by ϑ_H. Any single evaluation of $\vartheta_{H(\Delta)}\mathbf{f}(x)$ requires at most only eleven multiplications and ten additions. In fact, if $x \in [x_i, x_{i+1}]$, $0 \leq i \leq N$, then

$$\begin{aligned}\vartheta_{H(\Delta)}\mathbf{f}(x) = {} & f_i[\{(x_{i+1} - x_i)^{-1}(x - x_i)\}^2\{2(x_{i+1} - x_i)^{-1}(x - x_i) - 3\} + 1] \\ & + f_{i+1}[\{(x_{i+1} - x_i)^{-1}(x - x_i)\}^2\{-2(x_{i+1} - x_i)^{-1}(x - x_i) + 3\}] \\ & + f_i^1[\{(x_{i+1} - x_i)^{-1}(x - x_i)\}(x - x_i)(x_{i+1} - x)^2] \\ & + f_{i+1}^1[\{(x_{i+1} - x_i)^{-1}(x - x_i)\}^2(x - x_{i+1})],\end{aligned}$$

and we first compute $\theta \equiv (x_{i+1} - x_i)^{-1}(x - x_i)$, which requires two additions and one multiplication. Next we compute θ^2, which requires one multiplication, $2\theta - 3$, which requires one multiplication and one addition, $\theta^2(2\theta - 3)$, which requires one multiplication, and $\alpha \equiv \theta^2(x - x_{i+1})$, which requires one multiplication and one addition. Finally, we compute

$$\begin{aligned}\vartheta_{H(\Delta)}f(x) = f(x_i)[\theta^2(2\theta - 3) + 1] - f(x_{i+1})[\theta^2(2\theta - 3)] \\ + Df(x_i)[\alpha(x - x_{i+1})(x_{i+1} - x_i)] + Df(x_{i+1})[\alpha],\end{aligned}$$

which requires six multiplications and five additions.

We now examine how we can develop an interpolation scheme in $H(\Delta)$, which uses only the $N + 2$ values $\mathbf{f} \equiv (f_0, f_1, \ldots, f_{N+1})$ as does $\vartheta_{L(\Delta)}$. Our idea is to use local cubic Lagrange interpolation polynomials to approximate values, which we think of as the derivatives $f_i^1 \equiv Df(x_i)$, $0 \leq i \leq N + 1$, which in turn are used to compute an approximation to $\vartheta_{H(\Delta)}\mathbf{f}$.

More precisely, given $\{f_{k+i}\}_{i=0}^3$, $0 \leq k \leq N - 2$, we define

$$p_k(x) \equiv \sum_{i=0}^{3} \eta_{k,i}(x) f_{k+i},$$

where

$$\eta_{k,i}(x) \equiv \frac{\prod\limits_{\substack{j=0 \\ j\neq i}}^{3} (x - x_{k+j})}{\prod\limits_{\substack{j=0 \\ j\neq i}}^{3} (x_{k+i} - x_{k+j})},$$

which is the unique cubic polynomial interpolating $\{f_{k+i}\}_{i=0}^3$. If $N \geq 2$, i.e., if Δ has at least two interior points, we approximate the derivatives $f_i^1 \equiv Df(x_i)$, $0 \leq i \leq N + 1$, in the following fashion:

$$f_i^1 = Df(x_i) \approx \begin{cases} Dp_i(x_i), & i = 0, \\ Dp_{i-1}(x_i), & i = 1, \\ \frac{1}{2}(Dp_{i-2}(x_i) + Dp_{i-1}(x_i)), & 2 \leq i \leq N - 1, \\ Dp_{i-2}(x_i), & i = N, \text{ and} \\ Dp_{i-3}(x_i), & i = N + 1. \end{cases} \tag{3.1}$$

Using these approximations of the derivatives, we then compute the piecewise cubic Hermite interpolate as before, i.e.,

(3.2)
$$\begin{aligned}\tilde{\vartheta}_{H(\Delta)}\mathbf{f} \equiv \sum_{i=0}^{N+1} f_i h_i(x) &+ Dp_0(0)h_0^1(x) + Dp_0(x_1)h_1^1(x)\\ &+ \sum_{i=2}^{N-1} \tfrac{1}{2}(Dp_{i-2}(x_i) + Dp_{i-1}(x_i))h_i^1(x)\\ &+ Dp_{N-2}(x_N)h_N^1(x) + Dp_{N-2}(1)h_{N+1}^1(x).\end{aligned}$$

If $f_i \equiv f(x_i)$, $0 \leq i \leq N+1$, where $f(x)$ is a sufficiently smooth function, then we let $\tilde{\vartheta}_{H(\Delta)} f \equiv \tilde{\vartheta}_{H(\Delta)}\mathbf{f}$. Moreover, we can give a priori error bounds for the interpolation mapping $\tilde{\vartheta}_{H(\Delta)}$ in the same way as we will for the mapping $\vartheta_{H(\Delta)}$; cf. [3.9].

To get an idea of the behavior of the error in this procedure, we consider the simple function $f(x) \equiv x^4$. If $e_i(x) \equiv f(x) - \vartheta_H f(x)$, $x \in [x_i, x_{i+1}]$, $0 \leq i \leq N$, then we can verify by the Peano Kernel Theorem that

$$\max_{x \in [x_i, x_{i+1}]} |e_i(x)| = \tfrac{1}{384}(x_{i+1} - x_i)^4 4!, \qquad 0 \leq i \leq N.$$

Thus, $\| x^4 - \vartheta_{H(\Delta)} x^4 \|_\infty \leq 16h^4$, which shows that the piecewise cubic Hermite interpolation procedure is fourth-order accurate for $f(x) = x^4$. Using this bound we may compute the following table for *uniform* partitions $\Delta(h)$:

h	dim $H(\Delta(h))$	$\| x^4 - \vartheta_{H(\Delta(h))} x^4 \|_\infty$
1	4	0.16×10^2
10^{-1}	24	0.16×10^{-2}
10^{-2}	204	0.16×10^{-6}
10^{-3}	2004	0.16×10^{-10}

In Section 3.3, we will show that this special result generalizes and that this procedure is fourth-order accurate for all sufficiently smooth functions.

3.2 TWO-DIMENSIONAL PROBLEMS

In this section, we introduce a two-dimensional analogue of the interpolation procedure of the previous section.

We let $H(\rho) \equiv H(\Delta) \otimes H(\Delta_y)$ (the tensor product), i.e., $H(\rho)$ is the $4(N+2)(M+2)$-dimensional vector space of all functions of the form

$$\begin{aligned}h(x,y) = \sum_{i=0}^{N+1} \sum_{j=0}^{M+1} \{a_{ij}h_i(x)h_j(y) &+ b_{ij}h_i(x)h_j^1(y)\\ &+ c_{ij}h_i^1(x)h_j(y) + d_{ij}h_i^1(x)h_j^1(y)\}.\end{aligned}$$

Clearly $H(\rho)$ can be characterized as the vector space of all piecewise bicubic polynomials, $p(x, y)$, with respect to ρ, such that $D_x^l D_y^k p(x, y)$ is continuous on U for all $0 \leq l, k \leq 1$.

Given the vector

$$\mathbf{f} \equiv \{f_{ij}, f_{i,j}^{1,0}, f_{i,j}^{0,1}, f_{i,j}^{1,1}\}_{i=0, j=0}^{N+1, M+1}$$

we define

$$(3.3) \quad \vartheta_{H(\rho)}\mathbf{f} \equiv \sum_{i=0}^{N+1} \sum_{j=0}^{M+1} \{f_{ij}h_i(x)h_j(y) + f_{i,j}^{1,0}h_i^1(x)h_j(y) + f_{i,j}^{0,1}h_i(x)h_j^1(y) + f_{i,j}^{1,1}h_i^1(x)h_j^1(y)\}$$

as the interpolation mapping in $H(\rho)$. If $f_{ij} \equiv f(x_i, y_j)$, $f_{i,j}^{1,0} \equiv D_x f(x_i, y_j)$, $f_{i,j}^{0,1} \equiv D_y f(x_i, y_j)$, and $f_{i,j}^{1,1} \equiv D_x D_y f(x_i, y_j)$, for all $0 \leq i \leq N+1$ and $0 \leq j \leq M+1$, where $f(x, y)$, $D_x f(x, y)$, $D_y f(x, y)$, and $D_x D_y f(x, y)$ are defined for all $(x, y) \in U$, we will write $\vartheta_{H(\rho)} f$ for $\vartheta_{H(\rho)}\mathbf{f}$. We now give an important characterization of $\vartheta_{H(\rho)} f$ in terms of one-dimensional interpolation schemes.

THEOREM 3.1

If $f \in C^2(U)$, then

$$(3.4) \quad \vartheta_{H(\rho)} f = \vartheta_{H(\Delta_y)} \vartheta_{H(\Delta)} f = \vartheta_{H(\Delta)} \vartheta_{H(\Delta_y)} f.$$

Proof. We prove only the first equality in (3.4), since the second is proved the same way. By definition

$$\begin{aligned}\vartheta_{H(\Delta_y)} \vartheta_{H(\Delta)} f &= \vartheta_{H(\Delta_y)}\Big[\sum_{i=0}^{N+1} (f(x_i, y)h_i(x) + D_x f(x_i, y)h_i^1(x))\Big] \\ &= \sum_{j=0}^{M+1} \Big\{\Big[\sum_{i=0}^{N+1} (f(x_i, y_j)h_i(x) + D_x f(x_i, y_j)h_i^1(x))\Big]h_j(y) \\ &\quad + \Big[\sum_{i=0}^{N+1} (D_y f(x_i, y_j)h_i(x) + D_x D_y f(x_i, y_j)h_i^1(x))\Big]h_j^1(y)\Big\} \\ &= \vartheta_{H(\rho)} f. \qquad \text{Q.E.D.}\end{aligned}$$

Finally, following deBoor, we make some observations about the computational aspects of the two-dimensional problem; cf. [3.5]. By definition a piecewise bicubic polynomial is given by an expression of the form

$$(3.5) \quad c_{ij}(x, y) = \sum_{m,n=0}^{3} \gamma_{m,n}^{i,j}(x - x_{i-1})^m (y - y_{j-1})^n$$

in each subrectangle $U_{ij} \equiv \{(x, y) \mid x_{i-1} \leq x \leq x_i, y_{j-1} \leq y \leq y_j\}$, $1 \leq i \leq N+1$ and $1 \leq j \leq M+1$. Given $f(x, y)$, $D_x f(x, y)$, $D_y f(x, y)$, and $D_x D_y f(x, y)$ at the four corners of the rectangle U_{ij}, we know that there exists a unique bicubic polynomial $c_{ij}(x, y)$ which assumes the given values.

Moreover, the matrix of coefficients, $\Gamma_{ij} \equiv [\gamma^{ij}_{mn}]$, in (3.5) is given by

$$\Gamma_{ij} = A(\Delta x_{i-1})K_{ij}A^T(\Delta y_{j-1}), \tag{3.6}$$

where

$$K_{ij} \equiv \begin{bmatrix} B_{i-1,j-1} & B_{i-1,j} \\ B_{i,j-1} & B_{i,j} \end{bmatrix},$$

$$B_{l,k} \equiv \begin{bmatrix} f(x_l, y_k) & D_y f(x_l, y_k) \\ D_x f(x_l, y_k) & D_x D_y f(x_l, y_k) \end{bmatrix}, \text{ and}$$

$$A(h) \equiv \begin{bmatrix} 1 & 0 & 0 & 0 \\ 0 & 1 & 0 & 0 \\ -3h^{-2} & -2h^{-1} & 3h^{-2} & -h^{-1} \\ 2h^{-3} & h^{-2} & -2h^{-3} & h^{-2} \end{bmatrix}.$$

3.3 ERROR ANALYSIS

In this section, we prove a priori error bounds for the interpolation procedures introduced in Sections 3.1 and 3.2. In the one-dimensional case, our analysis is based upon the characterization of the piecewise cubic Hermite interpolate as the solution of a simple variational problem.

In fact, we will show that the piecewise cubic Hermite interpolate, $\vartheta_{H(\Delta)}\mathbf{f}$, is the Hermite interpolating function of minimum least squares curvature.

THEOREM 3.2

Let Δ and $\{f_i, f_i^1\}_{i=0}^{N+1}$ be given and

$$V \equiv \{w \in PC^{2,2}(I) \mid w(x_i) = f_i \text{ and } Dw(x_i) = f_i^1, 0 \leq i \leq N+1\}.$$

The variational problem of finding the functions $p \in V$ which minimize $\| D^2 w \|_2^2$ over all $w \in V$ has the unique solution $\vartheta_{H(\Delta)}\mathbf{f}$.

Proof. As in the proof of Theorem 2.2, $p \in V$ is a solution of the variational problem if and only if

$$(D^2 p, D^2 \delta)_2 = 0, \tag{3.7}$$

for all $\delta \in V_0 \equiv \{w \in PC^{2,2}(I) \mid w(x_i) = 0 \text{ and } Dw(x_i) = 0, 0 \leq i \leq N+1\}$, i.e., if and only if p is a solution of the generalized Euler equation. Moreover, the variational problem and the generalized Euler equation (3.7) have at most one solution.

We complete our proof by showing that $\vartheta_{H(\Delta)}\mathbf{f}$ is a solution of the generalized Euler equation (3.7). If $\delta \in V_0$, then integrating by parts, we have

$$
\begin{aligned}
(D^2\vartheta_{H(\Delta)}\mathbf{f}, D^2\delta)_2 &= \int_0^1 D^2\vartheta_{H(\Delta)}\mathbf{f}(x)D^2\delta(x)\,dx \\
&= \sum_{i=0}^{N} \int_{x_i}^{x_{i+1}} D^2\vartheta_{H(\Delta)}\mathbf{f}(x)D^2\delta(x)\,dx \\
&= \sum_{i=0}^{N} [D\delta(x)D^2\vartheta_{H(\Delta)}\mathbf{f}(x)]_{x_i}^{x_{i+1}} \\
&\quad - \sum_{i=0}^{N} \int_{x_i}^{x_{i+1}} D\delta(x)D^3\vartheta_{H(\Delta)}\mathbf{f}(x)\,dx \\
&= \sum_{i=0}^{N} [D\delta(x)D^2\vartheta_{H(\Delta)}\mathbf{f}(x)]_{x_i}^{x_{i+1}} \\
&\quad - \sum_{i=0}^{N} [\delta(x)D^3\vartheta_{H(\Delta)}\mathbf{f}(x)]_{x_i}^{x_{i+1}} \\
&\quad + \sum_{i=0}^{N} \int_{x_i}^{x_{i+1}} \delta(x)D^4\vartheta_{H(\Delta)}\mathbf{f}(x)\,dx.
\end{aligned}
$$

Thus, $(D^2\vartheta_{H(\Delta)}\mathbf{f}, D^2\delta)_2 = 0$, since the boundary terms vanish because of the interpolation conditions on δ, and each term of the last sum vanishes because $\vartheta_{H(\Delta)}\mathbf{f}$ is a cubic polynomial on each subinterval $[x_i, x_{i+1}]$, $0 \leq i \leq N$.

Q.E.D.

As a corollary of equation (3.7) of the preceding proof we obtain the so-called "First Integral Relation," which was originally introduced in [3.1].

COROLLARY

If $f \in PC^{2,2}(I)$, then

$$\| D^2\vartheta_H f \|_2^2 + \| D^2\vartheta_H f - D^2 f \|_2^2 = \| D^2 f \|_2^2. \tag{3.8}$$

By using the same type of integration by parts argument that we used in the proof of Theorem 3.2, we may prove the following result.

THEOREM 3.3

If $g \in PC^{4,2}(I)$, $g(x_i) = f_i$, and $Dg(x_i) = f_i^1$, $0 \leq i \leq N + 1$, then

$$\| D^2(g - \vartheta_H \mathbf{f}) \|_2^2 = (g - \vartheta_H \mathbf{f}, D^4 g)_2. \tag{3.9}$$

For the special case in which $f_i = f(x_i)$ and $f_i^1 = Df(x_i)$, $0 \leq i \leq N + 1$, and $g = f$, we obtain the so-called "Second Integral Relation," which was originally introduced in [3.1].

COROLLARY

If $f \in PC^{4,2}(I)$, then

$$\| D^2(f - \vartheta_H f)\|_2^2 = (f - \vartheta_H f, D^4 f)_2. \tag{3.10}$$

As in Chapter 2, we turn to the derivation of a priori bounds for the interpolation error, $f - \vartheta_H f$, and its derivatives with respect to the L^2-norm and the L^∞-norm. We find that if f is sufficiently smooth, then $\vartheta_H f$ is a fourth-order approximation to f with respect to both the L^∞-norm and the L^2-norm. We begin with a preliminary error bound in the L^2-norm which is not only of interest for its own sake, but will be used in the remainder of this chapter.

THEOREM 3.4

If $f \in PC^{2,2}(I)$, then

$$\| D^2(f - \vartheta_H f)\|_2 \leq \| D^2 f\|_2, \tag{3.11}$$

$$\| D(f - \vartheta_H f)\|_2 \leq \pi^{-1} h \| D^2 f\|_2, \tag{3.12}$$

and

$$\| f - \vartheta_H f\|_2 \leq \pi^{-2} h^2 \| D^2 f\|_2. \tag{3.13}$$

Proof. Inequality (3.11) follows directly from the First Integral Relation (3.8). To prove (3.12) we note that $Df(x_i) - D\vartheta_H f(x_i) = 0$, for all $0 \leq i \leq N + 1$, and by the Rayleigh–Ritz Inequality (Theorem 1.2)

$$\int_{x_i}^{x_{i+1}} [Df(x) - D\vartheta_H f(x)]^2 \, dx \leq \pi^{-2}(x_{i+1} - x_i)^2 \int_{x_i}^{x_{i+1}} [D^2 f(x) - D^2\vartheta_H f(x)]^2 \, dx \tag{3.14}$$

for all $0 \leq i \leq N$.

Summing both sides of inequality (3.14) with respect to i from 0 to N and taking the square root of both sides of the resulting inequality, we obtain

$$\| D(f - \vartheta_H f)\|_2 \leq \pi^{-1} h \| D^2(f - \vartheta_H f)\|_2, \tag{3.15}$$

and (3.12) follows by using (3.11) to bound the right-hand side of (3.15).

Inequality (3.13) is proved in an analogous way by using the fact that $f(x_i) - \vartheta_H f(x_i) = 0$, $0 \leq i \leq N$, and the Rayleigh–Ritz Inequality (Theorem 1.2) twice. Q.E.D.

If f is somewhat smoother, we may obtain stronger a priori bounds. The "boot strap" method of proof, which we will give, was suggested in [3.1] and refined in [3.8].

THEOREM 3.5

If $f \in PC^{4,2}(I)$, then

$$\| D^2(f - \vartheta_H f) \|_2 \leq \pi^{-2} h^2 \| D^4 f \|_2, \tag{3.16}$$

$$\| D(f - \vartheta_H f) \|_2 \leq \pi^{-3} h^3 \| D^4 f \|_2, \tag{3.17}$$

and

$$\| f - \vartheta_H f \|_2 \leq \pi^{-4} h^4 \| D^4 f \|_2. \tag{3.18}$$

Proof. Application of the Cauchy–Schwarz Inequality to the Second Integral Relation (3.10) yields

$$\| D^2(f - \vartheta_H f) \|_2^2 \leq \| D^4 f \|_2 \| f - \vartheta_H f \|_2. \tag{3.19}$$

Combining this with (3.13), we obtain (3.16). Now, using (3.16) to bound the right-hand side of (3.15), we obtain (3.17). Inequality (3.18) can be obtained in an analogous way. Q.E.D.

We now turn to a derivation of error bounds in the L^∞-norm.

THEOREM 3.6

If $f \in PC^{4,\infty}(I)$, then

$$\| f - \vartheta_H f \|_\infty \leq \frac{1}{384} h^4 \| D^4 f \|_\infty, \tag{3.20}$$

$$\| D(f - \vartheta_H f) \|_\infty \leq \frac{\sqrt{3}}{216} h^3 \| D^4 f \|_\infty, \tag{3.21}$$

$$\| D^2(f - \vartheta_H f) \|_\infty \leq \frac{1}{12} h^2 \| D^4 f \|_\infty, \tag{3.22}$$

and

$$\| D^3(f - \vartheta_H f) \|_\infty \leq \frac{1}{2} h \| D^4 f \|_\infty. \tag{3.23}$$

Proof. First we prove (3.20) and then we will indicate a proof of (3.21)–(3.23). For fixed $0 \leq i \leq N$, let $w(x) \equiv (x - x_i)(x - x_{i+1})$. For each $x \in [x_i, x_{i+1}]$, there exists $\xi_x \in [x_i, x_{i+1}]$ such that

$$e(x) \equiv f(x) - \vartheta_H f(x) = \frac{1}{4!} D^4 f(\xi_x) w^2(x).$$

In fact, if $x = x_i$ or x_{i+1}, any point ξ_x suffices. Otherwise, for fixed $\tilde{x}$, choose λ such that

$$\theta(\tilde{x}) \equiv e(\tilde{x}) - \lambda w^2(\tilde{x}) = 0.$$

But, counting multiplicities, $\theta(x)$ has five zeroes in $[x_i, x_{i+1}]$ and hence by Rolle's Theorem there exists a point $\xi_x \in [x_i, x_{i+1}]$ such that $D^4\theta(\xi_x) = 0$. But $D^4\theta(\xi_x) = D^4f(\xi_x) - 4!\lambda$ and hence $\lambda = \frac{1}{4!}D^4f(\xi_x)$. Thus,

$$\max_{x \in [x_i, x_{i+1}]} |e(x)| \leq \frac{1}{4!} \| D^4 f \|_\infty \max_{x \in [x_i, x_{i+1}]} |w^2(x)| \leq \frac{1}{384} h^4 \| D^4 f \|_\infty,$$

which proves (3.20).

To prove (3.21)–(3.23), the idea is to apply the Peano Kernel Theorem locally to the functional $D^j(I - \vartheta_H)$, $0 \leq j \leq 3$, on $PC^{4,\infty}(x_i, x_{i+1})$, for each $0 \leq i \leq N$. Doing this, we find that for all $x \in [x_i, x_{i+1}]$

$$D^j(I - \vartheta_H) f(x) = \int_{x_i}^{x_{i+1}} D_x^j K_x(t) D^4 f(t)\, dt,\ 0 \leq i \leq N,$$

where $K_x(t) \equiv (x - t)_+^4 - \vartheta_H(x - t)_+^4$. As may be easily verified, $K_x(t) \leq 0$ so that

$$\begin{aligned} \| (I - \vartheta_H) f(x) \|_\infty &\leq \| D^4 f \|_\infty \max_{0 \leq i \leq N} \int_{x_i}^{x_{i+1}} -K_x(t)\, dt \\ &\leq \max_{0 \leq i \leq N} \left\| \frac{1}{4!}(x - x_i)^2 (x - x_{i+1})^2 \right\|_\infty \| D^4 f \|_\infty \\ &\leq \frac{1}{384} h^4 \| D^4 f \|_\infty, \end{aligned}$$

which again proves (3.20).

However, it unfortunately turns out that $D_x^j K_x(t)$, $1 \leq j \leq 3$, has variable sign. Thus, to extend the preceding analysis we need to compute the zeroes of $D_x^j K_x(t)$, $1 \leq j \leq 3$, since the biggest possible error will occur for those functions, f, for which $| D^4 f(t) | = \| D^4 f \|_\infty$ for all $t \in [x_i, x_{i+1}]$, $0 \leq i \leq N$, and the sign of $D^4 f(t)$ is either always the same as that of $D_x^j K_x(t)$ or always the opposite. For a discussion of the computation of these zeroes and a completion of the derivation of the error bounds see [3.2]. Q.E.D.

To prove our next error bound, which will be used in the error analysis of the procedure for two-dimensional problems, we need the following preliminary result concerning the basis functions

$$\{h_i^1(x)\}_{i=0}^{N+1}.$$

LEMMA 3.1

$|h_i^1(x)| + |h_{i+1}^1(x)| \leq \frac{1}{4}h$ for all $x \in [x_i, x_{i+1}]$ and $0 \leq i \leq N$.

Proof. It clearly suffices to consider the interval $[0, h]$ and there

$$\eta(x) \equiv |h_i^1(x)| + |h_{i+1}^1(x)| = h^{-2}x(h - x)^2 + h^{-2}x^2(h - x).$$

To find the maximum of $\eta(x)$ on $[0, h]$, we set

$$D\eta(x) = h^{-2}[(h - x)^2 - 2x(h - x) + 2x(h - x) - x^2] = 0$$

to obtain $h^2 - 2xh = 0$. The root of this equation is $\tilde{x} = \frac{1}{2}h$ and in addition $D^2\eta(\tilde{x}) < 0$, which implies that $\tilde{x}$ maximizes η on $[0, h]$. Moreover, $\eta(\tilde{x}) = \frac{1}{4}$.

Q.E.D.

THEOREM 3.7

If $f \in PC^{2,\infty}(I)$, then

$$\|f - \vartheta_H f\|_\infty \leq \tfrac{1}{4}h^2 \|D^2 f\|_\infty. \tag{3.24}$$

Proof. For all $x \in [x_i, x_{i+1}]$, $0 \leq i \leq N$, we have

$$\begin{aligned} f(x) - \vartheta_H f(x) &= f(x) - \vartheta_L f(x) + \vartheta_L f(x) - \vartheta_H f(x) \\ &= f(x) - \vartheta_L f(x) + [D\vartheta_L f(x_i) - Df(x_i)]h_i^1(x) \\ &\quad + [D\vartheta_L f(x_{i+1}) - Df(x_{i+1})]h_{i+1}^1(x). \end{aligned}$$

Hence, using the results of Theorem 2.5 and Lemma 3.1, we have

$$\begin{aligned} \|f - \vartheta_H f\|_\infty &\leq \tfrac{1}{8}h^2 \|D^2 f\|_\infty + \|D\vartheta_L f - Df\|_\infty \max_{x \in [x_i, x_{i+1}]} (|h_i^1(x)| + |h_{i+1}^1(x)|) \\ &\leq \tfrac{1}{8}h^2 \|D^2 f\| + \tfrac{1}{8}h^2 \|D^2 f\|, \end{aligned}$$

which yields (3.24).

Q.E.D.

Our last error bound for our procedure for one-dimensional problems will be needed for the results about two-dimensional problems.

THEOREM 3.8

If $f \in PC^{3,2}(I)$, then

$$\|f - \vartheta_H f\|_2 \leq \pi^{-3}(1 + \pi^{-1}2\sqrt{15})h^3 \|D^3 f\|_2. \tag{3.25}$$

Proof. For fixed $0 \leq i \leq N$, let $q(x)$ be the unique quintic polynomial such that $D^j q(x_k) = D^j f(x_k)$ for $0 \leq j \leq 2$ and $k = i$ and $i + 1$. Then

$$\int_{x_i}^{x_{i+1}} (D^3 q(x))^2\, dx + \int_{x_i}^{x_{i+1}} (D^3 q(x) - D^3 f(x))^2\, dx = \int_{x_i}^{x_{i+1}} (D^3 f(x))^2\, dx. \tag{3.26}$$

In fact, by the Corollary to Theorem 1.4, it suffices to show that

$$\int_{x_i}^{x_{i+1}} (D^3q(x))(D^3q(x) - D^3f(x))\,dx = 0, \tag{3.27}$$

which can be verified by integration by parts coupled with the quintic interpolation conditions. Moreover, by the Rayleigh–Ritz Inequality (Theorem 1.2) we have

$$\int_{x_i}^{x_{i+1}} (q(x) - f(x))^2\,dx \leq \pi^{-6}(x_{i+1} - x_i)^6 \int_{x_i}^{x_{i+1}} (D^3f(x))^2\,dx. \tag{3.28}$$

Then, for all $x \in [x_i, x_{i+1}]$,

$$f(x) - \vartheta_H f(x) = f(x) - q(x) + q(x) - \vartheta_H q(x),$$

and hence by (3.18), (3.28), and the triangle and Schmidt Inequalities (cf. Theorem 1.5),

$$\begin{aligned}
&\left(\int_{x_i}^{x_{i+1}} (f(x) - \vartheta_H f(x))^2\,dx\right)^{1/2} \\
&\leq \left(\int_{x_i}^{x_{i+1}} (f(x) - q(x))^2\,dx\right)^{1/2} + \left(\int_{x_i}^{x_{i+1}} (q(x) - \vartheta_H q(x))^2\,dx\right)^{1/2} \\
&\leq \pi^{-3}(x_{i+1} - x_i)^3\left(\int_{x_i}^{x_{i+1}} (D^3f)^2\,dx\right)^{1/2} \\
&\qquad + \pi^{-4}(x_{i+1} - x_i)^4\left(\int_{x_i}^{x_{i+1}} (D^4q(x))^2\,dx\right)^{1/2} \\
&\leq \pi^{-3}(x_{i+1} - x_i)^3(1 + \pi^{-1}2\sqrt{15})\left(\int_{x_i}^{x_{i+1}} (D^3f(x))^2\,dx\right)^{1/2}.
\end{aligned} \tag{3.29}$$

The result follows by squaring both sides of the inequality (3.29), summing i from 0 to N, and taking the square root of both sides of the resulting inequality. Q.E.D.

We now proceed to the a priori error bounds for the piecewise bicubic Hermite interpolation procedure. As in the one-dimensional case, we find that if f is sufficiently smooth, then $\vartheta_{H(\rho)}f$ is a fourth-order approximation to f with respect to both the L^∞-norm and the L^2-norm.

THEOREM 3.9

If $f \in PC^{4,2}(U)$, then

$$\begin{aligned}
\| f - \vartheta_{H(\rho)}f \|_2 &\leq \pi^{-4}(h^4 \| D_x^4 f \|_2 + h^2k^2 \| D_x^2 D_y^2 f \|_2 + k^4 \| D_y^4 f \|_2) \\
&\leq \pi^{-4}\bar{\rho}^4(\| D_x^4 f \|_2 + \| D_x^2 D_y^2 f \|_2 + \| D_y^4 f \|_2),
\end{aligned} \tag{3.30}$$

$$(3.31)\quad \begin{aligned} \| D_x(f - \vartheta_{H(\rho)} f) \|_2 &\leq \pi^{-3} h^3 \| D_x^4 f \|_2 + \pi^{-3} h k^2 \| D_x^2 D_y^2 f \|_2 \\ &\quad + \pi^{-3}(1 + \pi^{-1} 2\sqrt{15}) k^3 \| D_y^3 D_x f \|_2 \\ &\leq \pi^{-3} \bar{\rho}^3 (\| D_x^4 f \|_2 + \| D_x^2 D_y^2 f \|_2 \\ &\quad + (1 + \pi^{-1} 2\sqrt{15}) \| D_y^3 D_x f \|_2), \end{aligned}$$

and

$$(3.32)\quad \begin{aligned} \| D_y(f - \vartheta_{H(\rho)} f) \|_2 &\leq \pi^{-3} k^3 \| D_y^4 f \|_2 + \pi^{-3} k h^2 \| D_x^2 D_y^2 f \|_2 \\ &\quad + \pi^{-3}(1 + \pi^{-1} 2\sqrt{15}) h^3 \| D_x^3 D_y f \|_2 \\ &\leq \pi^{-3} \bar{\rho}^3 (\| D_y^4 f \|_2 + \| D_x^2 D_y^2 f \|_2 \\ &\quad + (1 + \pi^{-1} 2\sqrt{15}) \| D_x^3 D_y f \|_2). \end{aligned}$$

Proof. From (3.4) and the triangle inequality, we have

$$(3.33)\quad \begin{aligned} \| f - \vartheta_{H(\rho)} f \|_2 &\leq \| f - \vartheta_{H(\Delta)} f \|_2 + \| \vartheta_{H(\Delta)}(f - \vartheta_{H(\Delta_y)} f) \|_2 \\ &\leq \| f - \vartheta_{H(\Delta)} f \|_2 + \| \vartheta_{H(\Delta)}(f - \vartheta_{H(\Delta_y)} f) \\ &\quad - (f - \vartheta_{H(\Delta_y)} f) \|_2 + \| f - \vartheta_{H(\Delta_y)} f \|_2. \end{aligned}$$

Using the results of Theorems 3.4 and 3.5 to bound the right-hand side of (3.33), we have

$$(3.34)\quad \begin{aligned} \| f - \vartheta_{H(\rho)} f \|_2 &\leq \pi^{-4} h^4 \| D_x^4 f \|_2 + \pi^{-2} h^2 \| D_x^2 (f - \vartheta_{H(\Delta_y)} f) \|_2 \\ &\quad + \pi^{-4} k^4 \| D_y^4 f \|_2. \end{aligned}$$

But since $D_x^2 \vartheta_{H(\Delta_y)} f = \vartheta_{H(\Delta_y)} D_x^2 f$, we have

$$(3.35)\quad \| D_x^2 (f - \vartheta_{H(\Delta_y)} f) \|_2 \leq \pi^{-2} k^2 \| D_x^2 D_y^2 f \|_2.$$

Using (3.35) to bound the right-hand side of (3.34), we obtain (3.30).

We may prove (3.31) and (3.32) in a similar way. For example, using the results of Theorems 3.4, 3.5, and 3.8, we have

$$\begin{aligned} \| D_x(f - \vartheta_{H(\rho)} f) \|_2 &\leq \| D_x(f - \vartheta_{H(\Delta)} f) \|_2 + \| D_x[\vartheta_{H(\Delta)}(f - \vartheta_{H(\Delta_y)} f) \\ &\quad - (f - \vartheta_{H(\Delta_y)} f)] \|_2 + \| D_x(f - \vartheta_{H(\Delta_y)} f) \|_2 \\ &\leq \pi^{-3} h^3 \| D_x^4 f \|_2 + \pi^{-1} h \| D_x^2 (f - \vartheta_{H(\Delta_y)} f) \|_2 \\ &\quad + \pi^{-3}(1 + \pi^{-1} 2\sqrt{15}) k^3 \| D_y^3 D_x f \|_2 \\ &\leq \pi^{-3} h^3 \| D_x^4 f \|_2 + \pi^{-3} h k^2 \| D_x^2 D_y^2 f \|_2 \\ &\quad + \pi^{-3}(1 + \pi^{-1} 2\sqrt{15}) k^3 \| D_y^3 f \|_2, \end{aligned}$$

which proves (3.31). Q.E.D.

The following error bounds for the L^∞-norm are proved in a similar way.

THEOREM 3.10

If $f \in PC^{4,\infty}(U)$, then

$$\begin{aligned} \| f - \vartheta_{H(\rho)} f \|_\infty &\le \tfrac{1}{384}(h^4 \| D_x^4 f \|_\infty + 24h^2k^2 \| D_x^2 D_y^2 f \|_\infty \\ &\qquad + k^4 \| D_y^4 f \|_\infty) \\ &\le \tfrac{1}{384}\bar{\rho}^4(\| D_x^4 f \|_\infty + 24 \| D_x^2 D_y^2 f \|_\infty + \| D_y^4 f \|_\infty). \end{aligned} \tag{3.36}$$

Proof. Again from (3.4) and the triangle inequality, we have

$$\begin{aligned} \| f - \vartheta_{H(\rho)} f \|_\infty &\le \| f - \vartheta_{H(\Delta)} f \|_\infty + \| \vartheta_{H(\Delta)}(f - \vartheta_{H(\Delta_y)} f) \|_\infty \\ &\le \| f - \vartheta_{H(\Delta)} f \|_\infty + \| \vartheta_{H(\Delta)}(f - \vartheta_{H(\Delta_y)} f) \\ &\qquad - (f - \vartheta_{H(\Delta_y)} f) \|_\infty + \| f - \vartheta_{H(\Delta_y)} f \|_\infty. \end{aligned} \tag{3.37}$$

Using the results of Theorems 3.6 and 3.7 to bound the right-hand side of (3.37), we have

$$\begin{aligned} \| f - \vartheta_{H(\rho)} f \|_\infty &\le \tfrac{1}{384} h^4 \| D_x^4 f \|_\infty + \tfrac{1}{4} h^2 \| D_x^2 (f - \vartheta_{H(\Delta_y)} f) \|_\infty \\ &\qquad + \tfrac{1}{384} k^4 \| D_y^4 f \|_\infty \\ &\le \tfrac{1}{384} h^4 \| D_x^4 f \|_\infty + \tfrac{1}{16} h^2 k^2 \| D_x^2 D_y^2 f \|_\infty + \tfrac{1}{384} k^4 \| D_y^4 f \|_\infty, \end{aligned}$$

which was to be proved. Q.E.D.

EXERCISES FOR CHAPTER 3

(3.1) Let

$$\Delta^1_{\alpha,N}: 0 < (N+1)^{-q} < \ldots < j^q (N+1)^{-q} < \ldots < (N+1)^q (N+1)^{-q} = 1,$$

where $q \equiv 4\alpha^{-1}$, be a partition of $[0, 1]$ for all $0 < \alpha < 4$, $\alpha \ne 1, 2, 3$. Show that if

$$E_j \equiv \max_{x \in [x_j, x_{j+1}]} | x^\alpha - \hat{\vartheta}_{H(\Delta^1_{\alpha,N})} x^\alpha |, \qquad 0 \le j \le N,$$

where we define

$$\hat{\vartheta}_{H(\Delta^1_{\alpha,N})} x^\alpha \equiv \begin{cases} \vartheta_{L(\Delta^1_{\alpha,N})} x^\alpha, & 0 \le x \le (N+1)^{-q}, \text{ and} \\ \vartheta_{H(\Delta^1_{\alpha,N})} x^\alpha, & (N+1)^{-q} \le x \le 1, \end{cases}$$

then

$$E_0 \leq (N+1)^{-4}$$

and

$$E_j \leq (q^4/384)[(j+1)/j]^{4(q-1)}(N+1)^{-4}\ \alpha|\alpha-1||\alpha-2||\alpha-3|, \quad 1 \leq j \leq N.$$

Thus, if we choose the partition properly, we can achieve a fourth-order accurate (in the L^∞-norm) interpolation scheme for x^α, $0 < \alpha < 4$, $\alpha \neq 1$, 2, 3, even though $x^\alpha \notin PC^{4,\infty}(I)$. Moreover,

$$\| x^\alpha - \hat{\vartheta}_{H(\Delta^1_{\alpha,N})} x^\alpha \|_\infty \leq \max\left(1, \frac{q^4}{384} 2^{4(q-1)} \alpha|\alpha-1||\alpha-2||\alpha-3|\right)(N+1)^{-4}$$

(cf. [3.7]).

(3.2) Use the Peano Kernel Theorem to show that if $f \in PC^{t,q}(I)$, $t = 1, 2, 3$, or 4, then there exists a positive constant, K, which can be explicitly computed, such that

$$\| f - \vartheta_H f \|_p \leq \begin{cases} K h^{t+p^{-1}-q^{-1}} \| D^t f \|_q, & \text{if } p \geq q \geq 1, \\ K h^t \| D^t f \|_q, & \text{if } q \geq p \geq 1, \end{cases}$$

for all partitions Δ of I.

(3.3) Use Exercises (1.2) and (1.4) to show that if $f \in PC^{2,2}(I)$, then for all $p \geq 2$,

$$\| D(f - \vartheta_H f) \|_p \leq \tfrac{1}{2} h^{1/2+p^{-1}} \| D^2 f \|_2$$

and

$$\| f - \vartheta_H f \|_p \leq (2\pi)^{-1} h^{3/2+p^{-1}} \| D^2 f \|_2.$$

Furthermore, show that if $f \in PC^{4,2}(I)$, then for all $p \geq 2$,

$$\| D(f - \vartheta_H f) \|_p \leq (2\pi^2)^{-1} h^{5/2+p^{-1}} \| D^4 f \|_2$$

and

$$\| f - \vartheta_H f \|_p \leq (2\pi^3)^{-1} h^{7/2+p^{-1}} \| D^4 f \|_2.$$

(3.4) Show that if $f \in PC^{2,2}(I)$, then for all $p \leq 2$,

$$\| D(f - \vartheta_H f) \|_p \leq \pi^{-1} h \| D^2 f \|_2$$

and

$$\| f - \vartheta_H f \|_p \leq \pi^{-2} h^2 \| D^2 f \|_2.$$

Furthermore, show that if $f \in PC^{4,2}(I)$, then for all $p \leq 2$,

$$\| D(f - \vartheta_H f) \|_p \leq \pi^{-3} h^3 \| D^4 f \|_2$$

and

$$\| f - \vartheta_H f \|_p \leq \pi^{-4} h^4 \| D^4 f \|_2.$$

(3.5) Show that if $f \in PC^{2,2}(I)$, then $\vartheta_H f$ satisfies the "local First Integral Relation,"

$$\int_{x_i}^{x_{i+1}} (D^2\vartheta_H f(x))^2\,dx + \int_{x_i}^{x_{i+1}} (D^2\vartheta_H f(x) - D^2 f(x))^2\,dx = \int_{x_i}^{x_{i+1}} (D^2 f(x))^2\,dx, \qquad 0 \le i \le N.$$

(3.6) Show that if $f \in PC^{4,2}(I)$, then $\vartheta_H f$ satisfies the "local Second Integral Relation,"

$$\int_{x_i}^{x_{i+1}} (D^2 f(x) - D^2\vartheta_H f(x))^2\,dx = \int_{x_i}^{x_{i+1}} (f(x) - \vartheta_H f(x))D^4 f(x)\,dx, \qquad 0 \le i \le N.$$

(3.7) Using the results of Exercises (3.5) and (3.6), prove local versions of the results of Theorems 3.4 and 3.5.

(3.8) For each positive integer, m, and each partition, Δ, of I, let $H^m(\Delta)$ be the vector space of all piecewise polynomials, p, of degree $2m - 1$ with respect to Δ such that $p \in C^{m-1}(I)$. Show that $H^1(\Delta) = L(\Delta)$ and $H^2(\Delta) = H(\Delta)$. Moreover, given

$$\mathbf{f}^m \equiv (f_0^0, f_0^1, \ldots, f_0^{m-1}, f_1^0, \ldots, f_{N+1}^0, f_{N+1}^1, \ldots, f_{N+1}^{m-1}),$$

let $\vartheta_{H^m(\Delta)}\mathbf{f}^m$ be the unique element, $h(x)$, in $H^m(\Delta)$ such that

$$D^j h(x_i) = f_i^j, \qquad 0 \le i \le N+1, \quad 0 \le j \le m-1.$$

Show that the mapping $\vartheta_{H^m(\Delta)}$ is well-defined for all $m \ge 1$.

(3.9) Using the notations of Exercise (3.8), show that if $p \in PC^{m,2}(I)$ and $D^j p(x_i) = 0$, $0 \le i \le N+1$, $0 \le j \le m-1$, then $(D^m p, D^m h)_2 = 0$ for all $h \in H^m(\Delta)$. Furthermore, given $\mathbf{f}^m$, show that $\vartheta_{H^m(\Delta)}\mathbf{f}^m$ is the unique solution of the variational problem

$$\inf\{\| D^m p \|_2 \mid p \in PC^{m,2}(I),\ D^j p(x_i) = f_i^j,\ 0 \le i \le N+1,\ 0 \le j \le m-1\},$$

and if $f \in PC^{m,2}(I)$ and $\vartheta_{H^m} f \equiv \vartheta_{H^m}\mathbf{f}^m$, where $f_i^j \equiv D^j f(x_i)$, $0 \le i \le N+1$, $0 \le j \le m-1$, prove the "First Integral Relation,"

$$\| D^m\vartheta_{H^m} f \|_2^2 + \| D^m\vartheta_{H^m} f - D^m f \|_2^2 = \| D^m f \|_2^2.$$

(3.10) Using the notations of Exercises (3.8) and (3.9), prove the "Second Integral Relation" for $f \in PC^{2m,2}(I)$, i.e.,

$$\| D^m(f - \vartheta_{H^m} f) \|_2^2 = (f - \vartheta_{H^m} f, D^{2m} f)_2.$$

(3.11) Using the notations and results of Exercises (3.8), (3.9), and (3.10), show that if $f \in PC^{m,2}(I)$, then

$$\| D^j(f - \vartheta_{H^m} f) \|_2 \le \pi^{m-j} h^{m-j} \| D^m f \|_2, \qquad 0 \le j \le m,$$

and if $f \in PC^{2m,2}(I)$, then

$$\| D^j(f - \vartheta_{H^m} f) \|_2 \leq \pi^{2m-j} h^{2m-j} \| D^{2m} f \|_2.$$

(3.12) Use the Peano Kernel Theorem to show that if $f \in PC^{t,q}(I)$, $1 \leq t \leq 2m$, then there exists a positive constant, K, which can be explicitly computed, such that

$$\| f - \vartheta_{H^m} f \|_p \leq \begin{cases} K h^{t+p^{-1}-q^{-1}} \| D^t f \|_q, & \text{if } p \geq q \geq 1, \\ K h^t \| D^t f \|_q, & \text{if } q \geq p \geq 1, \end{cases}$$

for all partitions Δ of I (cf. [3.3]).

(3.13) Use the results of Exercises (1.2), (1.4), and (3.11) to show that if $f \in PC^{m,2}(I)$, then for all $p \geq 2$

$$\| D^j(f - \vartheta_{H^m} f) \|_p \leq \tfrac{1}{2} \pi^{-m+1+j} h^{m-1/2+p^{-1}-j} \| D^m f \|_2, \quad 0 \leq j \leq m-1.$$

Furthermore, if $f \in PC^{2m,2}(I)$, then for all $p \geq 2$

$$\| D^j(f - \vartheta_{H^m} f) \|_p \leq \tfrac{1}{2} \pi^{-2m+1+j} h^{2m-1/2+p^{-1}-j} \| D^{2m} f \|_2, \quad 0 \leq j \leq m-1.$$

(3.14) Show that if $f \in PC^{m,2}(I)$, then for all $p \leq 2$

$$\| D^j(f - \vartheta_{H^m} f) \|_p \leq \pi^{-m+j} h^{m-j} \| D^m f \|_2,$$

and if $f \in PC^{2m,2}(I)$, then for all $p \leq 2$

$$\| D^j(f - \vartheta_{H^m} f) \|_p \leq \pi^{-2m+j} h^{2m-j} \| D^{2m} f \|_2.$$

(3.15) Show that if $f \in PC^{m,2}(I)$, then $\vartheta_{H^m} f$ satisfies the "local First Integral Relation,"

$$\int_{x_i}^{x_{i+1}} (D^m \vartheta_{H^m} f(x))^2 \, dx + \int_{x_i}^{x_{i+1}} (D^m \vartheta_{H^m} f(x) - D^m f(x))^2 \, dx = \int_{x_i}^{x_{i+1}} (D^m f(x))^2 \, dx, \qquad 0 \leq i \leq N.$$

(3.16) Show that if $f \in PC^{2m,2}(I)$, then $\vartheta_{H^m} f$ satisfies the "local Second Integral Relation,"

$$\int_{x_i}^{x_{i+1}} (D^m f(x) - D^m \vartheta_{H^m} f(x))^2 \, dx = \int_{x_i}^{x_{i+1}} (f(x) - \vartheta_{H^m} f(x)) D^{2m} f(x) \, dx, \qquad 0 \leq i \leq N.$$

(3.17) Using the results of Exercises (3.15) and (3.16), prove local versions of the results of Exercise (3.11).

(3.18) Let $\vartheta_{H^m(\rho)} \equiv \vartheta_{H^m(\Delta)} \vartheta_{H^m(\Delta_y)}$. Using the results of Exercise (3.11), show that if $f \in PC^{2m,2}(U)$, then

$$\| f - \vartheta_{H^m(\rho)} f \|_2$$
$$\leq \pi^{-2m} h^{2m} \| D_x^{2m} f \|_2 + \pi^{-2m} h^m k^m \| D_x^m D_y^m f \|_2 + \pi^{-2m} k^{2m} \| D_y^{2m} f \|_2$$

(cf. [3.3], [3.4]).

(3.19) Let $\vartheta_{H^m(\rho)} \equiv \vartheta_{H^m(\Delta)} \vartheta_{H^m(\Delta_y)}$. Using the results of Exercise (3.12), show that if $f \in PC^{2m,\infty}(U)$, then there exists a positive constant, K, such that

$$\| f - \vartheta_{H^m(\rho)} f \|_\infty \leq K(h^{2m} \| D_x^{2m} f \|_\infty + h^m k^m \| D_x^m D_y^m f \|_\infty + k^{2m} \| D_y^{2m} f \|_\infty)$$

(cf. [3.3], [3.4]).

(3.20) Show that if $f \in PC^{p,2}(I)$, where $m < p < 2m$, then

$$\| D^j(f - \vartheta_{H^m} f) \|_2$$
$$\leq \pi^{-p+j} \{1 + \pi^{p-2m} 2^{1/2(2m-p)} p! [(2p - 2m)!]^{-1}\} h^{p-j} \| D^p f \|_2,$$

for all $0 \leq j \leq m$.

REFERENCES FOR CHAPTER 3

[3.1] AHLBERG, J. H., E. N. NILSON, and J. L. WALSH, Convergence properties of generalized splines. *Proc. Nat. Acad. Sci. U.S.A.* **54**, 344–350 (1965).

[3.2] BIRKHOFF, G., and A. PRIVER, Hermite interpolation errors for derivatives. *J. Math. and Physics* **46**, 440–447 (1967).

[3.3] BIRKHOFF, G., M. H. SCHULTZ, and R. S. VARGA, Piecewise Hermite interpolation in one and two variables with applications to partial differential equations. *Numer. Math.* **11**, 232–256 (1968).

[3.4] BRAMBLE, J., and S. HILBERT, Bounds for a class of linear functionals with application to Hermite interpolation. *Numer. Math.* **16**, 362–369 (1971).

[3.5] DEBOOR, C., Bicubic spline interpolation. *J. Math. and Physics* **41**, 212–218 (1962).

[3.6] HALL, C. A., Bicubic interpolation over triangles. *J. Math. and Mech.* **19**, 1–11 (1969).

[3.7] RICE, John R., On the Degree of Convergence of Nonlinear Spline Approximation. *Approximations with Special Emphasis on Spline Functions* (I. J. Schoenberg, ed.) 349–367, Academic Press, New York (1969).

[3.8] SCHULTZ, M. H., and R. S. VARGA, *L*-splines. *Numer. Math.* **10**, 345–369 (1967).

[3.9] SCHULTZ, M. H., *Computing piecewise Hermite interpolates without derivatives.* Yale Computer Science Research Report.

4 SPLINE INTERPOLATION

4.1 ONE-DIMENSIONAL PROBLEMS

In this chapter, we introduce and study a (cubic) spline interpolation procedure which is fourth-order accurate. Our spline interpolation procedure is an improvement over the piecewise cubic Hermite interpolation procedure of Chapter 3 in the sense that it yields a smoother interpolate, i.e., the spline interpolate is a $C^2(I)$-function while the piecewise cubic Hermite interpolate is only a $C^1(I)$-function. Moreover, the spline interpolate depends on roughly half as many parameters as the piecewise cubic Hermite interpolate. We begin with the basic definition of (cubic) splines due originally to Schoenberg; cf. [4.9].

DEFINITION 4.1

Given Δ, let the *space of cubic splines with respect to* Δ, $S(\Delta)$, be the vector space of all twice continuously differentiable, piecewise cubic polynomials on I with respect to Δ, i.e.,

$$S(\Delta) \equiv \{p(x) \in C^2(I) \mid p(x) \text{ is a cubic polynomial on each subinterval } [x_i, x_{i+1}],\ 0 \leq i \leq N, \text{ defined by } \Delta\}.$$

Clearly a C^1-piecewise cubic Hermite polynomial $h(x)$, i.e., $h(x) \in H(\Delta)$, is a spline function if and only if $h \in C^2(I)$. Conversely, every cubic spline function $s(x) \in H(\Delta)$ and we have the inclusion $S(\Delta) \subset H(\Delta)$. Moreover, the dimension of $S(\Delta)$ is $N + 4$, while the dimension of $H(\Delta)$ is $2N + 4$. Thus we may represent every cubic spline function $s(x)$ in terms of the basis functions $\{h_i(x), h_i^1(x)\}_{i=0}^{N+1}$ of $H(\Delta)$ even though these basis functions do *not* belong to $S(\Delta)$. In fact, we have

$$s(x) = \sum_{i=0}^{N+1} (s(x_i)h_i(x) + Ds(x_i)h_i^1(x)).$$

In this book we will use the words cubic spline and spline interchangeably, although, as we will see in the exercises, the concept of spline function can be greatly generalized.

It is natural to define the following cubic spline interpolation procedure. Ahlberg, Nilson, and Walsh (cf. [4.1]) refer to this procedure as the Type I procedure. See the exercises and the references given there for a discussion of interpolation procedures of Types II, III, and IV.

DEFINITION 4.2

Given $\mathbf{f} \equiv \{f_0, \ldots, f_{N+1}, f_0^1, f_{N+1}^1\} \in R^{N+3}$, let $\vartheta_{S(\Delta)}\mathbf{f}$, the *$S(\Delta)$-interpolate of* $\mathbf{f}$, be the unique spline, $s(x)$, in $S(\Delta)$ such that $s(x_i) = f_i$, $0 \leq i \leq N+1$, and $Ds(x_i) = f_i^1$, $i = 0$ and $N+1$.

We now show that this procedure is well-defined. Following deBoor (cf. [4.5]), we first prove the following result.

LEMMA 4.1

Let $x_{i-1} < x_i < x_{i+1}$ for some $1 \leq i \leq N$ and $p(x)$ and $q(x)$ be two cubic polynomials such that $p(x_i) = q(x_i) = y_i$ and $Dp(x_i) = Dq(x_i) = y_i^1$. Then $D^2p(x_i) = D^2q(x_i)$ if and only if

$$\begin{aligned} &\Delta x_i Dp(x_{i-1}) + 2(\Delta x_i + \Delta x_{i-1})y_i^1 + \Delta x_{i-1} Dq(x_{i+1}) \\ &\quad = 3[\Delta x_{i-1}(\Delta x_i)^{-1}(q(x_{i+1}) - y_i) + \Delta x_i(\Delta x_{i-1})^{-1}(y_i - p(x_{i-1}))], \end{aligned} \tag{4.1}$$

where $\Delta x_{i-1} \equiv x_i - x_{i-1}$ and $\Delta x_i \equiv x_{i+1} - x_i$.

Proof. Clearly,

$$\begin{aligned} p(x) = p(x_i) &+ Dp(x_i)(x - x_i) \\ &+ [3(p(x_{i-1}) - p(x_i))(x_i - x_{i-1})^{-2} \\ &+ (Dp(x_{i-1}) + 2Dp(x_i))(x_i - x_{i-1})^{-1}](x - x_i)^2 \\ &+ [2(p(x_{i-1}) - p(x_i))(x_i - x_{i-1})^{-3} \\ &+ (Dp(x_{i-1}) + Dp(x_i))(x_i - x_{i-1})^{-2}](x - x_i)^3, \end{aligned}$$

and hence

$$\begin{aligned} D^2p(x_i) = 2[3(p(x_{i-1}) - p(x_i))(x_i - x_{i-1})^{-2} \\ + (Dp(x_{i-1}) + 2Dp(x_i))(x_i - x_{i-1})^{-1}]. \end{aligned}$$

Similarly,

$$\begin{aligned}q(x) = q(x_i) + Dq(x_i)(x - x_i) + [3(q(x_{i+1}) - q(x_i))(x_i - x_{i+1})^{-2} \\ + (Dq(x_{i+1}) + 2Dq(x_i))(x_i - x_{i+1})^{-1}](x - x_i)^2 \\ + [2(q(x_{i+1}) - q(x_i))(x_i - x_{i+1})^{-3} \\ + (Dq(x_{i+1}) + Dq(x_i))(x_i - x_{i+1})^{-2}](x - x_i)^3,\end{aligned}$$

and hence

$$\begin{aligned}D^2q(x_i) = 2[3(q(x_{i+1}) - q(x_i))(x_i - x_{i+1})^{-2} \\ + (Dq(x_{i+1}) + 2Dq(x_i))(x_i - x_{i+1})^{-1}].\end{aligned}$$

Thus, we have equality if and only if

$$\begin{aligned}3(p(x_{i-1}) - p(x_i))(\Delta x_{i-1})^{-2} + (Dp(x_{i-1}) + 2Dp(x_i))(\Delta x_{i-1})^{-1} \\ = 3(q(x_{i+1}) - q(x_i))(\Delta x_i)^{-2} - (Dq(x_{i+1}) + 2Dq(x_i))(\Delta x_i)^{-1}.\end{aligned}$$

Q.E.D.

We can now prove that the spline interpolation procedure is well-defined.

THEOREM 4.1

Let $h(x) \in H(\Delta)$. For given numbers $c_i \equiv h(x_i)$, $0 \leq i \leq N + 1$, and $c_0^1 \equiv Dh(0)$, $c_{N+1}^1 \equiv Dh(1)$, there exists exactly one set of given numbers $c_i^1 \equiv Dh(x_i)$, $1 \leq i \leq N$, such that $h \in S(\Delta)$.

Proof. By Lemma 4.1 the continuity of $D^2h(x)$ is equivalent to the set of N linear equations

$$\begin{aligned}\Delta x_i c_{i-1}^1 + 2(\Delta x_i + \Delta x_{i-1})c_i^1 + \Delta x_{i-1} c_{i+1}^1 \\ = 3[\Delta x_{i-1}(\Delta x_i)^{-1}\Delta c_i + \Delta x_i(\Delta x_{i-1})^{-1}\Delta c_{i-1}],\end{aligned} \tag{4.2}$$

$1 \leq i \leq N$, where $\Delta x_j \equiv x_{j+1} - x_j$ and $\Delta c_j \equiv c_{j+1} - c_j$, $0 \leq j \leq N$, for the N unknown numbers c_i^1, $1 \leq i \leq N$. The linear equations (4.2) can be rewritten in vector form as

$$B\mathbf{c}^1 = \mathbf{k}, \tag{4.3}$$

where $B \equiv [b_{ij}]$,

$$b_{ij} \equiv \begin{cases} 2(\Delta x_i + \Delta x_{i-1}), & 1 \leq j = i \leq N, \\ \Delta x_i, & 1 \leq j = i - 1 \leq N - 1, \\ \Delta x_{i-1}, & 2 \leq j = i + 1 \leq N, \\ 0, & \text{otherwise,} \end{cases} \tag{4.4}$$

i.e., B is a tridiagonal matrix, and $\mathbf{k} \equiv [k_i]$,

$$(4.5) \qquad k_i \equiv \begin{cases} 3[\Delta x_0(\Delta x_1)^{-1}\Delta c_1 + \Delta x_1(\Delta x_0)^{-1}\Delta c_0] - \Delta x_1 Dh(0), & i = 1, \\ 3[\Delta x_{i-1}(\Delta x_i)^{-1}\Delta c_i + \Delta x_i(\Delta x_{i-1})^{-1}\Delta c_{i-1}], & 1 < i < N, \\ 3[\Delta x_{N-1}(\Delta x_N)^{-1}\Delta c_N + \Delta x_N(\Delta x_{N-1})^{-1}\Delta c_{N-1}] - \Delta x_{N-1} Dh(1), & \\ & i = N. \end{cases}$$

The tridiagonal matrix, B, of the linear system is strictly diagonally dominant and hence by the Gerschgorin Theorem (cf. [4.7]), the system has a unique solution. Q.E.D.

If $f_i \equiv f(x_i)$, $0 \leq i \leq N+1$, and $f_i^1 \equiv Df(x_i)$, $i = 0$ and $N+1$, where $f \in C^1(I)$, then we will denote $\vartheta_{S(\Delta)}\mathbf{f}$ by $\vartheta_{S(\Delta)}f$. Moreover, we will usually omit the Δ from both notations.

We compute $\vartheta_{S(\Delta)}\mathbf{f}$ by solving the system (4.3) for $\mathbf{c}^1$ and writing

$$\vartheta_{S(\Delta)}\mathbf{f} = \sum_{i=0}^{N+1} f_i h_i(x) + f_0^1 h_0^1(x) + \sum_{i=1}^{N} c_i^1 h_i^1(x) + f_{N+1}^1 h_{N+1}^1(x).$$

To construct the matrix B, we first compute and store Δx_i, $0 \leq i \leq N$, which requires $N+1$ arithmetic operations, and then we compute $2(\Delta x_i + \Delta x_{i-1})$, $1 \leq i \leq N$, which requires $2N$ operations, for a total of $3N+1$ operations, where we count both additions and multiplications as operations. To construct the vector $\mathbf{k}$, we first compute $\Delta c_i \equiv f_{i+1} - f_i$, $0 \leq i \leq N$, which requires $N+1$ operations, and then we compute the expressions given in (4.5) which requires $6N+4$ operations, for a total of $7N + 5$ operations. Finally, solving the linear system (4.3) by Gaussian elimination for tridiagonal matrices requires $5N - 1$ operations. Thus, it requires a total of $15N + 5$ arithmetic operations to compute the vector $\mathbf{c}^1$. Furthermore, it follows from the results of Section 3.1 that with this type of representation of $\vartheta_{S(\Delta)}\mathbf{f}$, a single evaluation of $\vartheta_{S(\Delta)}\mathbf{f}(x)$ requires at most only twenty arithmetic operations.

We now examine how we can develop an interpolation mapping $\tilde{\vartheta}_{S(\Delta)}$ in $S(\Delta)$, which uses only the $N+2$ values $\mathbf{f} \equiv (f_0, f_1, \ldots, f_{N+1})$ as does $\vartheta_{L(\Delta)}$. As in Chapter 3, our idea is to use the derivative of the local cubic Lagrange interpolating polynomial at both ends of the interval to approximate values, which we think of as the derivatives $f_0^1 = Df(0)$ and $f_{N+1}^1 = Df(1)$, which in turn are used to compute an approximation to $\vartheta_{S(\Delta)}\mathbf{f}$.

More precisely, if $N \geq 2$, i.e., if Δ has at least two interior points, we define

$$p_0(x) \equiv \sum_{i=0}^{3} \eta_{0,i}(x) f_i$$

and

$$p_{N-2}(x) \equiv \sum_{i=0}^{3} \eta_{N-2,i}(x) f_{N-2+i},$$

where

$$\eta_{k,i}(x) \equiv \frac{\prod_{\substack{j=0\\j\neq 1}}^{3}(x - x_{k+j})}{\prod_{\substack{j=0\\j\neq 1}}^{3}(x_{k+i} - x_{k+j})}$$

for $k = 0$ and $N - 2$. We approximate the derivative $f_0^1 \equiv Df(0)$ and $f_{N+1}^1 \equiv Df(1)$ in the following fashion:

$$Df(0) \approx Dp_0(0) \quad \text{and} \quad Df(1) \approx Dp_{N-2}(1). \tag{4.6}$$

Using these approximations of the derivatives, we then compute the cubic spline interpolate as before.

If $f_i \equiv f(x_i)$, $0 \leq i \leq N + 1$, where $f(x)$ is a sufficiently smooth function, then we let $\tilde{\vartheta}_{S(\Delta)} f \equiv \tilde{\vartheta}_{S(\Delta)}\mathbf{f}$, and we can give a priori error bounds for the interpolation mapping $\tilde{\vartheta}_{S(\Delta)}$ in the same way as we will in Section 4.3 for the mapping $\vartheta_{S(\Delta)}$; cf. [4.12].

Using the result of Theorem 4.1, it is possible to describe a basis for $S(\Delta)$, namely the "cardinal splines," $\{C_i(x)\}_{i=0}^{N+3}$, defined by the following interpolation conditions:

$$\begin{aligned} &C_j(x_i) = \delta_{ij}, \qquad DC_j(0) = DC_j(1) = 0, \qquad 0 \leq i, j \leq N + 1,\\ &C_{N+2}(x_i) = C_{N+3}(x_i) = 0, \qquad 0 \leq i \leq N + 1,\\ &DC_{N+2}(0) = DC_{N+3}(1) = 1, \text{ and } DC_{N+2}(1) = DC_{N+2}(0) = 0. \end{aligned}$$

Clearly,

$$\vartheta_S\mathbf{f}(x) = \sum_{i=0}^{N+1} f_i C_i(x) + f_0^1 C_{N+2}(x) + f_{N+1}^1 C_{N+3}(x)$$

and

$$s(x) = \sum_{i=0}^{N+1} s(x_i)C_i(x) + Ds(0)C_{N+2}(x) + Ds(1)C_{N+3}(x)$$

for all $s \in S(\Delta)$. Moreover, the interpolation mapping $\vartheta_S\mathbf{f}$ is *not* local, i.e., $\vartheta_S\mathbf{f}(x)$ depends on *all* the quantities f_i, $0 \leq i \leq N + 1$, f_0^1, and f_{N+1}^1.

To get an idea of the behavior of the error in this procedure, we consider the simple function $f(x) = e^x$. For the special case of $x_0 = 0$, $x_1 = \frac{1}{2}$, and $x_2 = 1$, the linear system (4.3) reduces to one linear equation in one unknown, c_1^1. Its solution is $c_1^1 = -\frac{1}{4}(1 + e) + \frac{3}{4}(e - 1) = \frac{1}{2}e - 1$, and $\vartheta_{S(0,1/2,1)} e^x = h_0(x) + h_0^1(x) + e^{1/2}h_1(x) + (\frac{1}{2}e - 1)h_1^1(x) + eh_2(x) + eh_2^1(x)$.

For uniform partitions $\Delta(h)$ with more knots, we have computed the following table:

h	dim $S(\Delta(h))$	$\|\| e^x - \vartheta_{S(\Delta(h))} e^x \|\|_\infty$
$\frac{1}{4}$	7	0.26×10^{-4}
$\frac{1}{5}$	8	0.11×10^{-4}
$\frac{1}{6}$	9	0.53×10^{-5}
$\frac{1}{7}$	10	0.29×10^{-5}
$\frac{1}{8}$	11	0.17×10^{-5}

Since the error apparently decreases by a factor of approximately $16 = 2^4$ when we halve the mesh length, we have that the spline interpolation procedure is fourth-order accurate for $f(x) = e^x$. In Section 4.3 we will prove that this special result generalizes and that this procedure is fourth-order accurate for all sufficiently smooth functions.

4.2 TWO-DIMENSIONAL PROBLEMS

In this section, we introduce a two-dimensional analogue of the interpolation procedure of the previous section. We consider only the case of rectangular partitions of $U \equiv [0, 1] \times [0, 1]$. See [4.3] for nontrivial results for more general domains in the plane.

We let $S(\rho) \equiv H(\Delta) \otimes H(\Delta_y)$, i.e., $S(\rho)$ is the $(N + 4)(M + 4)$-dimensional vector space of all functions of the form

$$s(x, y) = \sum_{i=0}^{N+3} \sum_{j=0}^{M+3} \beta_{ij} C_i(x) C_j(y).$$

Clearly $S(\rho)$ is the space of all $C^2(U)$, piecewise bicubic polynomials with respect to ρ.

Given the vector

$$\mathbf{f} \equiv \{f_{ij}, f_{0,j}^{1,0}, f_{N+1,j}^{1,0}, f_{i,0}^{0,1}, f_{i,M+1}^{0,1}, f_{0,0}^{1,1}, f_{0,M+1}^{1,1}, f_{N+1,0}^{1,1}, f_{N+1,M+1}^{1,1}\}_{i=0,\, j=0}^{N+1,\, M+1},$$

we define

$$\begin{aligned} \vartheta_{S(\rho)} \mathbf{f} = & \sum_{i=0}^{N+1} \sum_{j=0}^{M+1} f_{i,j} C_i(x) C_j(y) \\ & + \sum_{j=0}^{M+1} (f_{0,j}^{1,0} C_{N+2}(x) + f_{N+1,j}^{1,0} C_{N+3}(x)) C_j(y) \\ & + \sum_{i=0}^{N+1} (f_{i,0}^{0,1} C_{M+2}(y) + f_{i,M+1}^{0,1} C_{M+3}(y)) C_i(x) \\ & + f_{0,0}^{1,1} C_{N+2}(x) C_{M+2}(y) + f_{0,M+1}^{1,1} C_{N+2}(x) C_{M+3}(y) \\ & + f_{N+1,0}^{1,1} C_{N+3}(x) C_{M+2}(y) + f_{N+1,M+1}^{1,1} C_{N+3}(x) C_{M+3}(y) \end{aligned} \tag{4.7}$$

as the interpolation mapping in $S(\rho)$. If $f_{i,j} \equiv f(x_i, y_j)$, $f_{0,j}^{1,0} \equiv D_x f(0, y_j)$, $f_{N+1,j}^{1,0} \equiv D_x f(1, y_j)$, $f_{i,0}^{0,1} \equiv D_y f(x_i, 0)$, $f_{i,M+1}^{0,1} \equiv D_y f(x_i, 1)$, $f_{0,0}^{1,1} \equiv D_x D_y f(0, 0)$, $f_{0,M+1}^{1,1} \equiv D_x D_y f(0, 1)$, $f_{N+1,0}^{1,1} \equiv D_x D_y f(1, 0)$, and $f_{N+1,M+1}^{1,1} \equiv D_x D_y f(1, 1)$ for all $0 \leq i \leq N + 1$ and $0 \leq j \leq M + 1$, where $f(x, y)$, $D_x f(x, y)$, $D_y f(x, y)$, and $D_x D_y f(x, y)$ are defined for all $(x, y) \in U$, we will write $\vartheta_{S(\rho)} f$ for $\vartheta_{S(\rho)}\mathbf{f}$. We now give an important characterization of $\vartheta_{S(\rho)} f$ in terms of one-dimensional interpolation schemes.

THEOREM 4.2

If $f \in C^2(U)$, then

$$\vartheta_{S(\rho)} f = \vartheta_{S(\Delta_y)} \vartheta_{S(\Delta)} f = \vartheta_{S(\Delta)} \vartheta_{S(\Delta_y)} f. \tag{4.8}$$

Proof. We prove only the first equality in (4.8), as the second is proved the same way. By definition,

$$\begin{aligned}
\vartheta_{S(\Delta_y)} \vartheta_{S(\Delta)} f &= \vartheta_{S(\Delta_y)}\left[\sum_{i=0}^{N+1} f(x_i, y) C_i(x) + D_x f(0, y) C_{N+2}(x)\right. \\
&\qquad \left. + D_x f(1, y) C_{N+3}(x)\right] \\
&= \sum_{j=0}^{M+1}\left[\sum_{i=0}^{N+1} f(x_i, y_j) C_i(x) + D_x f(0, y_j) C_{N+2}(x)\right. \\
&\qquad \left. + D_x f(1, y_j) C_{N+3}(x)\right] C_j(y) \\
&\qquad + \left[\sum_{i=0}^{N+1} D_y f(x_i, 0) C_i(x) + D_x D_y f(0, 0) C_{N+2}(x)\right. \\
&\qquad \left. + D_x D_y f(1, 0) C_{N+3}(x)\right] C_{N+2}(y) + \left[\sum_{i=0}^{N+1} D_y f(x_i, 1) C_i(x)\right. \\
&\qquad \left. + D_x D_y f(0, 1) C_{N+2}(x) + D_x D_y f(1, 1) C_{N+3}(x)\right] C_{N+3}(y) \\
&= \vartheta_{S(\rho)} f.
\end{aligned}$$

Q.E.D.

Finally, we comment on the computation of $\vartheta_{S(\rho)} f$. Since the restriction of $\vartheta_{S(\rho)} f$ to each vertical and horizontal line of the mesh ρ is a one-dimensional cubic spline, we may use the given data to do one-dimensional spline interpolation along these lines. This procedure gives us the values of $D_x \vartheta_{S(\rho)} f(x_i, y_j)$ and $D_y \vartheta_{S(\rho)} f(x_i, y_j)$ for all mesh points (x_i, y_j), $0 \leq i \leq N + 1$ and $0 \leq j \leq M + 1$.

To obtain the mixed partial derivatives $D_x D_y \vartheta_{S(\rho)} f(x_i, y_j)$, we first note that the restriction of $D_x \vartheta_{S(\rho)} f(x, y)$ to the vertical boundaries of U is a spline in y, and we have just the right data to do one-dimensional spline interpolation there. This yields the mixed partial derivatives at all the points of the form (x_i, y_j), $i = 0$ and $N + 1$, $0 \leq j \leq M + 1$. Second, to obtain the mixed partial derivatives at the other partition points, we note that the restriction of

$D_y \vartheta_{S(\rho)} f(x, y)$ to the horizontal lines in ρ is a spline in x, and moreover, we now have just the right data to do a one-dimensional spline interpolation there.

If $f_{i,j} \equiv f(x_i, y_j)$, $f_{i,j}^{1,0} \equiv D_x f(x_i, y_j)$, $f_{i,j}^{0,1} \equiv D_y f(x_i, y_j)$, and $f_{i,j}^{1,1} \equiv D_x D_y f(x_i, y_j)$, $0 \leq i \leq N+1$ and $0 \leq j \leq M+1$, then the preceding procedure is equivalent to solving the following $2M + N + 8$ sets of linear equations:

$$\begin{aligned} &\Delta x_{i-1} f_{i+1,j}^{1,0} + 2(\Delta x_{i-1} + \Delta x_i) f_{i,j}^{1,0} + \Delta x_i f_{i-1,j}^{1,0} \\ &\quad = 3[\Delta x_{i-1}(\Delta x_i)^{-1}(f_{i+1,j} - f_{i,j}) + \Delta x_i(\Delta x_{i-1})^{-1}(f_{i,j} - f_{i-1,j})], \\ &\qquad\qquad 1 \leq i \leq N,\ 0 \leq j \leq M+1, \end{aligned} \tag{4.9}$$

$$\begin{aligned} &\Delta y_{j-1} f_{i,j+1}^{0,1} + 2(\Delta y_{j-1} + \Delta y_j) f_{i,j}^{0,1} + \Delta y_j f_{i,j-1}^{0,1} \\ &\quad = 3[\Delta y_{j-1}(\Delta y_j)^{-1}(f_{i,j+1} - f_{i,j}) + \Delta y_j(\Delta y_{j-1})^{-1}(f_{i,j} - f_{i,j-1})], \\ &\qquad\qquad 1 \leq j \leq M, \quad 0 \leq i \leq N+1, \end{aligned} \tag{4.10}$$

$$\begin{aligned} &\Delta y_{j-1} f_{i,j+1}^{1,1} + 2(\Delta y_{j-1} + \Delta y_j) f_{i,j}^{1,1} + \Delta y_j f_{i,j-1}^{1,1} \\ &\quad = 3[\Delta y_{j-1}(\Delta y_j)^{-1}(f_{i,j+1}^{1,0} - f_{i,j}^{1,0}) + \Delta y_j(\Delta y_{j-1})^{-1}(f_{i,j}^{1,0} - f_{i,j-1}^{1,0})], \\ &\qquad\qquad 1 \leq j \leq M, \quad i = 0 \text{ and } N+1, \end{aligned} \tag{4.11}$$

and

$$\begin{aligned} &\Delta x_{i-1} f_{i+1,j}^{1,1} + 2(\Delta x_{i-1} + \Delta x_i) f_{i,j}^{1,1} + \Delta x_i f_{i-1,j}^{1,1} \\ &\quad = 3[\Delta x_{i-1}(\Delta x_i)^{-1}(f_{i+1,j}^{0,1} - f_{i,j}^{0,1}) + \Delta x_i(\Delta x_{i-1})^{-1}(f_{i,j}^{0,1} - f_{i-1,j}^{0,1})], \\ &\qquad\qquad 1 \leq i \leq N, \quad 0 \leq j \leq M+1. \end{aligned} \tag{4.12}$$

4.3 ERROR ANALYSIS

In this section, we prove a priori error bounds for the interpolation procedures introduced in Sections 4.1 and 4.2. In the one-dimensional case, our analysis is based upon the fact that the spline interpolating function describes the shape of a thin beam passing through the interpolation points and "clamped" at the end points. As such, it can be characterized as the solution of a simple variational problem.

We now state and prove this variational characterization of $\vartheta_{S(\Delta)}\mathbf{f}$.

THEOREM 4.3

Let Δ and $\mathbf{f} \equiv \{f_0, \ldots, f_{N+1}, f_0^1, f_{N+1}^1\}$ be given and $V \equiv \{w \in PC^{2,2}(I) \mid w(x_i) = f_i,\ 0 \leq i \leq N+1 \text{ and } Dw(x_i) = f_i^1,\ i = 0 \text{ and } N+1\}$. The variational problem of finding the functions $p \in V$ which minimize $\| D^2 w \|_2^2$ over all $w \in V$ has the unique solution $\vartheta_{S(\Delta)}\mathbf{f}$.

Proof. As in the proof of Theorem 2.2, $p \in V$ is a solution of the variational problem if and only if

$$(D^2 p, D^2 \delta)_2 = 0, \tag{4.13}$$

for all $\delta \in V_0 \equiv \{w \in PC^{2,2}(I) \mid w(x_i) = 0, 0 \leq i \leq N+1, \text{and } Dw(x_i) = 0, i = 0 \text{ and } N+1\}$, i.e., if and only if p is a solution of the generalized Euler equation. Moreover, the variational problem and the generalized Euler equation (4.13) have at most one solution.

We complete our proof by showing that $\vartheta_{S(\Delta)}\mathbf{f}$ is a solution of the generalized Euler equation (4.13). If $\delta \in V_0$, then integrating by parts twice we have

$$\begin{aligned}
(D^2\vartheta_{S(\Delta)}\mathbf{f}, D^2\delta)_2 &= \int_0^1 D^2\vartheta_{S(\Delta)}\mathbf{f}(x) D^2\delta(x)\,dx \\
&= \sum_{i=0}^{N} \int_{x_i}^{x_{i+1}} D^2\vartheta_{S(\Delta)}\mathbf{f}(x) D^2\delta(x)\,dx \\
&= \sum_{i=0}^{N} [D\delta(x) D^2\vartheta_{S(\Delta)}\mathbf{f}(x)]_{x_i}^{x_{i+1}} \\
&\quad - \sum_{i=0}^{N} [\delta(x) D^3\vartheta_{S(\Delta)}\mathbf{f}(x)]_{x_i}^{x_{i+1}} \\
&\quad + \sum_{i=0}^{N} \int_{x_i}^{x_{i+1}} \delta(x) D^4\vartheta_{S(\Delta)}\mathbf{f}(x)\,dx.
\end{aligned} \tag{4.14}$$

Thus, $(D^2\vartheta_{S(\Delta)}\mathbf{f}, D^2\delta)_2 = 0$ since the first sum on the right-hand side of (4.14) is equal to zero because of the smoothness of $\delta(x)$ and $\vartheta_{S(\Delta)}\mathbf{f}(x)$ and of the fact that $D\delta(0) = D\delta(1) = 0$; each term of the second sum is equal to zero because $\delta(x_i) = 0$, $0 \leq i \leq N+1$; and finally each term of the last sum is equal to zero because $D^4\vartheta_{S(\Delta)}\mathbf{f}(x) = 0$ for all $x \in [x_i, x_{i+1}]$, $0 \leq i \leq N$. Q.E.D.

As a corollary to the preceding proof we obtain the so-called "First Integral Relation" for splines, which was originally introduced in [4.1].

COROLLARY

If $f \in PC^{2,2}(I)$, then

$$\| D^2\vartheta_S f \|_2^2 + \| D^2\vartheta_S f - D^2 f \|_2^2 = \| D^2 f \|_2^2. \tag{4.15}$$

By using the same type of integration by parts argument that we used in the proof of Theorem 4.3, we may prove the following result.

THEOREM 4.4

If $g \in PC^{4,2}(I)$, $g(x_i) = f_i$, $0 \leq i \leq N + 1$, and $Dg(x_i) = f_i^1$, $i = 0$ and $N + 1$, then

$$\| D^2(g - \vartheta_S \mathbf{f}) \|_2^2 = (g - \vartheta_S \mathbf{f}, D^4 g)_2. \tag{4.16}$$

For the special case in which $f_i = f(x_i)$, $0 \leq i \leq N + 1$, and $f_i^1 = Df(x_i)$, $i = 0$ and $N + 1$, and $g = f$, we obtain the so-called "Second Integral Relation" for splines, which was originally introduced in [4.1].

COROLLARY

If $f \in PC^{4,2}(I)$, then

$$\| D^2(f - \vartheta_S f) \|_2^2 = (f - \vartheta_S f, D^4 f)_2. \tag{4.17}$$

As in Chapters 2 and 3, we now turn to the derivation of a priori bounds for the interpolation error, $f - \vartheta_S f$, and its derivatives with respect to the L^2-norm and the L^∞-norm. We find as in Chapter 3 that if f is sufficiently smooth, then $\vartheta_S f$ is a fourth-order approximation to f with respect to both the L^∞-norm and the L^2-norm. We start with a preliminary result, which is of interest for its own sake.

THEOREM 4.5

If $f \in PC^{2,2}(I)$, then

$$\| D^2(f - \vartheta_S f) \|_2 \leq \| D^2 f \|_2, \tag{4.18}$$

$$\| D(f - \vartheta_S f) \|_2 \leq 2\pi^{-1} h \| D^2 f \|_2, \tag{4.19}$$

and

$$\| f - \vartheta_S f \|_2 \leq 2\pi^{-2} h^2 \| D^2 f \|_2. \tag{4.20}$$

Proof. Inequality (4.18) follows directly from the First Integral Relation (4.15). To prove (4.19), we note that by Rolle's Theorem applied to $e(x) \equiv f(x) - \vartheta_S f(x)$, there exist points $\{\xi_l\}_{l=0}^N$ in $[0, 1]$ such that $x_k < \xi_k < x_{k+1}$, $0 \leq k \leq N$, and $De(\xi_k) = 0$, $0 \leq k \leq N$.

Thus, applying the Rayleigh–Ritz Inequality (Theorem 1.2), we have

$$\int_{\xi_k}^{\xi_{k+1}} [De(x)]^2 \, dx \leq \pi^{-2}(2h)^2 \int_{\xi_k}^{\xi_{k+1}} [D^2 e(x)]^2 \, dx, \qquad 0 \leq k \leq N - 1, \tag{4.21}$$

$$\int_0^{\xi_0} [De(x)]^2 \, dx \leq \pi^{-2} h^2 \int_0^{\xi_0} [D^2 e(x)]^2 \, dx, \tag{4.22}$$

and

$$\int_{\xi_N}^1 [De(x)]^2 \, dx \leq \pi^{-2} h^2 \int_{\xi_N}^1 [D^2 e(x)]^2 \, dx. \tag{4.23}$$

Summing both sides of inequality (4.21) with respect to k from 0 to $N-1$, adding inequalities (4.22) and (4.23) to the resulting inequality, and taking the square root of both sides of the resulting inequality, we obtain

$$\| D(f - \vartheta_S f) \|_2 \leq 2\pi^{-1} h \| D^2(f - \vartheta_S f) \|_2, \tag{4.24}$$

and (4.19) follows by using (4.18) to bound the right-hand side of (4.24).

Inequality (4.20) follows in a similar fashion from

$$\| e \|_2 \leq \pi^{-1} h \| De \|_2 \tag{4.25}$$

and (4.19). Q.E.D.

If f is somewhat smoother, we may obtain stronger a priori bounds, as we did in Chapters 2 and 3. The "boot strap" method of proof, which we will use again, was suggested in [4.1] and refined in [4.10].

THEOREM 4.6

If $f \in PC^{4,2}(I)$, then

$$\| D^2(f - \vartheta_S f) \|_2 \leq 2\pi^{-2} h^2 \| D^4 f \|_2, \tag{4.26}$$

$$\| D(f - \vartheta_S f) \|_2 \leq 4\pi^{-3} h^3 \| D^4 f \|_2, \tag{4.27}$$

and

$$\| f - \vartheta_S f \|_2 \leq 4\pi^{-4} h^4 \| D^4 f \|_2. \tag{4.28}$$

Proof. Applying the Cauchy–Schwarz Inequality to the Second Integral Relation, we obtain

$$\| D^2(f - \vartheta_S f) \|_2^2 \leq \| D^4 f \|_2 \| f - \vartheta_S f \|_2. \tag{4.29}$$

Combining this with (4.20) yields (4.26).

Now, using (4.26) to bound the right-hand side of (4.24), we obtain (4.27). Finally, using (4.27) to bound the right-hand side of (4.25), we obtain (4.28). Q.E.D.

To obtain an a priori error bound in the L^∞-norm, we follow Hall (cf. [4.6]), and write

$$f - \vartheta_S f = (f - \vartheta_H f) + (\vartheta_H f - \vartheta_S f) \tag{4.30}$$

and use the results of Chapter 3 to bound the first term on the right-hand side. To bound the second term, we observe that $\vartheta_H f - \vartheta_S f \in H(\Delta)$, $(\vartheta_H f - \vartheta_S f)(x_i) = 0$, $0 \leq i \leq N+1$, and $(D\vartheta_H f - \vartheta_S f)(x_i) = 0$, $i = 0$ and $N+1$. Hence,

$$\vartheta_H f(x) - \vartheta_S f(x) = \sum_{i=1}^{N} (D\vartheta_H f(x_i) - D\vartheta_S f(x_i))h_i^1(x)$$
$$= \sum_{i=1}^{N} (Df(x_i) - D\vartheta_S f(x_i))h_i^1(x)$$
$$= \sum_{i=1}^{N} e_i^1 h_i^1(x),$$

and it suffices to obtain an a priori bound on the vector $\mathbf{e}^1 \equiv [e_i^1]$.

LEMMA 4.2

If $f \in PC^{4,\infty}(I)$, then

$$\|\mathbf{e}^1\|_\infty \equiv \max_{1\leq i\leq N} |e_i^1| \leq \tfrac{1}{24}h^3\|D^4 f\|_\infty. \tag{4.31}$$

If $f \in C^5(I)$ and Δ is a uniform partition of I, then

$$\|\mathbf{e}^1\|_\infty \leq \tfrac{1}{60}h^4\|D^5 f\|_\infty. \tag{4.32}$$

Proof. The proof of (4.31) is accomplished by showing via the Peano Kernel Theorem that if $f_i^1 \equiv Df(x_i)$, $1 \leq i \leq N$, and $\mathbf{f}^1 \equiv [f_i^1] \in R^N$, then $B\mathbf{f}^1 = \mathbf{k} + \mathbf{r}$ and hence $B\mathbf{e}^1 = \mathbf{r}$. Inequality (4.31) is then obtained from a priori bounds on the norms of B^{-1} and $\mathbf{r} \equiv [r_i]$.

For each $1 \leq i \leq N$, we have from (4.2) that

$$\begin{aligned} r_i(f) = {} & \Delta x_i Df(x_{i-1}) + 2(\Delta x_i + \Delta x_{i-1})Df(x_i) + \Delta x_{i-1}Df(x_{i+1}) \\ & - 3[\Delta x_i(\Delta x_{i-1})^{-1}(f(x_i) - f(x_{i-1})) \\ & + \Delta x_{i-1}(\Delta x_i)^{-1}(f(x_{i+1}) - f(x_i))], \end{aligned} \tag{4.33}$$

and, by the Peano Kernel Theorem, we have

$$r_i(f) = \tfrac{1}{6}\int_{x_{i-1}}^{x_{i+1}} D^4 f(t)(r_i)_x(x-t)_+^3\,dt. \tag{4.34}$$

It can easily be verified by direct computation that

$$(r_i)_x(x-t)_+^3 = \begin{cases} 6(\Delta x_i + \Delta x_{i-1})(x_i - t)^2 + 3\Delta x_{i-1}(x_{i+1} - t)^2 \\ -3[\Delta x_i(\Delta x_{i-1})^{-1}(x_i - t)^3 \\ \quad + \Delta x_{i-1}(\Delta x_i)^{-1}\{x_{i+1} - t)^3 - (x_i - t)^3\}], \\ \qquad\qquad x_{i-1} \leq t \leq x_i, \text{ and} \\ 3\Delta x_{i-1}(x_{i+1} - t)^2 - 3[\Delta x_{i-1}(\Delta x_i)^{-1}(x_{i+1} - t)^3], \\ \qquad\qquad x_i \leq t \leq x_{i+1}. \end{cases} \tag{4.35}$$

Moreover, $(r_i)_x(x-t)_+^3 \geq 0$ for all $t \in [x_i, x_{i+1}]$ and $(r_i)_x(x-t)_+^3 \leq 0$ for all $t \in [x_{i-1}, x_i]$. In fact, for $t \in [x_i, x_{i+1}]$,

$$(r_i)_x(x-t)_+^3 = 3\Delta x_{i-1}(x_{i+1}-t)^2[1-(x_{i+1}-t)(x_{i+1}-x_i)^{-1}] \geq 0,$$

while for $t \in [x_{i-1}, x_i]$, $(r_i)_x(x-t)_+^3$ is a cubic polynomial, $q(t)$, such that

$$q(x_i) = 3\Delta x_{i-1}(\Delta x_i)^2 - 3\Delta x_{i-1}(\Delta x_i)^{-1}(\Delta x_i)^3 = 0,$$

$$\begin{aligned} q(x_{i-1}) &= 6(\Delta x_i + \Delta x_{i-1})(\Delta x_{i-1})^2 + 3\Delta x_{i-1}(\Delta x_i + \Delta x_{i-1})^2 \\ &\quad - 3[\Delta x_i(\Delta x_{i-1})^{-1}(\Delta x_{i-1})^3 + \Delta x_{i-1}(\Delta x_i)^{-1}\{(\Delta x_i + \Delta x_{i-1})^3 \\ &\quad - (\Delta x_{i-1})^3\}] \\ &= 0, \end{aligned}$$

$$Dq(x_i) = -6\Delta x_{i-1}(\Delta x_i) + 9\Delta x_{i-1}(\Delta x_i)^{-1}(\Delta x_i)^2 > 0,$$

and

$$\begin{aligned} Dq(x_{i-1}) &= -12(\Delta x_i + \Delta x_{i-1})(\Delta x_{i-1}) - 6\Delta x_{i-1}(\Delta x_i + \Delta x_{i-1}) \\ &\quad + 9[\Delta x_i(\Delta x_{i-1})^{-1}(\Delta x_{i-1})^2 + \Delta x_{i-1}(\Delta x_i)^{-1}\{(\Delta x_i)^2 \\ &\quad + 2(\Delta x_i)(\Delta x_{i-1})\}] \\ &= 0. \end{aligned}$$

Thus, the cubic polynomial, $q(t)$, vanishes at $t = x_{i-1}$ and x_i and $Dq(t)$ vanishes at $t = x_{i-1}$ and is positive at x_i. Hence, $q(t) \leq 0$.

Using these facts on the sign of $(r_i)_x(x-t)_+^3$, we have from (4.34) that

$$\begin{aligned} |r_i(f)| &\leq \frac{\|D^4 f\|_\infty}{6}\left\{\int_{x_{i-1}}^{x_i} -(r_i)_x(x-t)_+^3\,dt + \int_{x_i}^{x_{i+1}} (r_i)_x(x-t)_+^3\,dt\right\} \\ &= \frac{\|D^4 f\|_\infty}{6}\{2(\Delta x_i + \Delta x_{i-1})(x_i - t)^3\,|_{x_{i-1}}^{x_i} + \Delta x_{i-1}(x_{i+1}-t)^3\,|_{x_{i-1}}^{x_i} \\ &\quad - \tfrac{3}{4}[\Delta x_i(\Delta x_{i-1})^{-1}(x_i - t)^4 + \Delta x_{i-1}(\Delta x_i)^{-1}\{(x_{i+1}-t)^4 \\ &\quad - (x_i - t)^4\}]\,|_{x_{i-1}}^{x_i} - \Delta x_{i-1}(x_{i+1}-t)^3\,|_{x_i}^{x_{i+1}} \\ &\quad + \tfrac{3}{4}\Delta x_{i-1}(\Delta x_i)^{-1}(x_{i+1}-t)^4\,|_{x_i}^{x_{i+1}}\} \\ &\leq \tfrac{1}{24}\|D^4 f\|_\infty[\Delta x_i(\Delta x_{i-1})^3 + \Delta x_{i-1}(\Delta x_i)^3]. \end{aligned} \tag{4.36}$$

Multiplying both sides of $B\mathbf{e}^1 = \mathbf{r}$ by the diagonal matrix $D \equiv [d_{ij}]$, where $d_{ii} \equiv \frac{1}{2}(\Delta x_i + \Delta x_{i-1})^{-1}$, $1 \leq i \leq N$, we have

$$DB\mathbf{e}^1 = (I + M)\mathbf{e}^1 = D\mathbf{r}$$

and

$$\|M\|_\infty \equiv \|[m_{ij}]\|_\infty \equiv \max_{1 \leq j \leq N} \sum_{i=1}^{N} |m_{ij}| = \tfrac{1}{2}.$$

Thus,

$$\| (DB)^{-1} \|_\infty = \| (I + M)^{-1} \|_\infty \le 2$$

(cf. [4.7]), and

$$\| \mathbf{e}^1 \|_\infty \le 2 \| D\mathbf{r} \|_\infty. \tag{4.37}$$

However, from (4.36) we have

$$\begin{aligned} |d_{ii} r_i| &\le \tfrac{1}{48} \| D^4 f \|_\infty [\Delta x_i (\Delta x_{i-1})^3 + \Delta x_{i-1} (\Delta x_i)^3] (\Delta x_i + \Delta x_{i-1})^{-1} \\ &\le \tfrac{1}{48} \| D^4 f \|_\infty [\max (\Delta x_{i-1}, \Delta x_i)]^3. \end{aligned}$$

In fact, if for example $0 < \Delta x_{i-1} \le \Delta x_i$, then

$$\begin{aligned} &[\Delta x_i (\Delta x_{i-1})^3 + \Delta x_{i-1} (\Delta x_i)^3] (\Delta x_i + \Delta x_{i-1})^{-1} \\ &\quad \le \Delta x_i \Delta x_{i-1} [(\Delta x_i)^2 + (\Delta x_i)^2] (\Delta x_{i-1} + \Delta x_{i-1})^{-1} \\ &\quad = (\Delta x_i)^3. \end{aligned}$$

Hence

$$\| D\mathbf{r} \|_\infty \le \tfrac{1}{48} h^3 \| D^4 f \|_\infty,$$

and using this to bound the right-hand side of (4.37) we obtain (4.31).

For the case of a uniform partition, i.e., $\Delta x_i = h$, $0 \le i \le N$, we have from (4.33) that

$$\begin{aligned} \tfrac{1}{3} r_i(f) &= \tfrac{1}{3} h [Df(x_{i-1}) + 2Df(x_i) + Df(x_{i+1})] - [f(x_{i+1}) - f(x_{i-1})] \\ &= \tfrac{1}{3} h [Df(x_{i-1}) + 2Df(x_i) + Df(x_{i+1})] - \int_{x_{i-1}}^{x_{i+1}} Df(x)\, dx, \\ &\qquad 1 \le i \le N. \end{aligned} \tag{4.38}$$

The right-hand side of (4.38) turns out by coincidence to be exactly the error for Simpson's rule for the approximate calculation of the integral $\int_{x_{i-1}}^{x_{i+1}} Df(x)\, dx$, and it is a classical result of numerical analysis that

$$|\tfrac{1}{3} r_i(f)| \le \frac{h^5}{90} \| D^4(Df) \|_\infty, \qquad 1 \le i \le N;$$

cf. [4.7]. Using the a priori bound $\| \mathbf{r} \|_\infty \le (h^5/30) \| D^5 f \|_\infty$, we may complete the proof as in the nonuniform case to obtain $\| D\mathbf{r} \| \le (h^4/120) \| D^5 f \|_\infty$ and (4.32). Q.E.D.

THEOREM 4.7

If $f \in PC^{4,\infty}(I)$, then

$$\| f - \vartheta_S f \|_\infty \le \tfrac{5}{384} h^4 \| D^4 f \|_\infty. \tag{4.39}$$

Moreover, if $f \in C^5(I)$ and Δ is a uniform partition, then

$$\|f - \vartheta_S f\|_\infty \leq h^4(\tfrac{1}{384}\|D^4 f\|_\infty + \tfrac{1}{240}h\|D^5 f\|_\infty). \tag{4.40}$$

Proof. As we previously noted,

$$\|f - \vartheta_S f\|_\infty \leq \|f - \vartheta_H f\|_\infty + \|\vartheta_H f - \vartheta_S f\|_\infty.$$

Using the bound (3.20) and Lemma 3.1 in conjunction with Lemma 4.2, we have in the general case

$$\|f - \vartheta_S f\|_\infty \leq \tfrac{1}{384}h^4\|D^4 f\|_\infty + \tfrac{1}{96}h^4\|D^4 f\|_\infty = \tfrac{5}{384}h^4\|D^4 f\|_\infty.$$

In the case of $f \in C^5(I)$ and a uniform partition, we have

$$\|f - \vartheta_S f\|_\infty \leq \tfrac{1}{384}h^4\|D^4 f\|_\infty + \tfrac{1}{60}h^5\|D^5 f\|_\infty,$$

which yields (4.40). Q.E.D.

Birkhoff and deBoor gave the first proof of the fact that cubic spline interpolation was fourth-order accurate in the L^∞-norm, at least for partitions with bounded mesh ratios; cf. [4.2]. However, our treatment follows that of Hall, who proved the result without any restrictions; cf. [4.6].

Our last result of this section is of interest not only for itself but for its usefulness in obtaining error bounds for bicubic spline interpolation.

THEOREM 4.8

If $f \in PC^{2,\infty}(I)$, then

$$\|f - \vartheta_S f\|_\infty \leq \tfrac{2}{3}h^2\|D^2 f\|_\infty. \tag{4.41}$$

Proof. The proof is essentially the same as that of Theorem 4.7. The only difference lies in the bound for $\|D\mathbf{r}\|_\infty$.

By the Peano Kernel Theorem,

$$r_i(f) = \int_{x_{i-1}}^{x_{i+1}} D^2 f(t)(r_i)_x(x-t)_+\,dt, \qquad 1 \leq i \leq N,$$

and

$$(r_i)_x(x-t)_+ = \begin{cases} 2(\Delta x_i + \Delta x_{i-1}) + \Delta x_{i-1} - 3[\Delta x_i(\Delta x_{i-1})^{-1}(x_i - t) \\ \quad + \Delta x_{i-1}(\Delta x_i)^{-1}\{(x_{i+1} - t) - (x_i - t)\}], & x_{i-1} \leq t \leq x_i, \\ \Delta x_{i-1} - 3[\Delta x_{i-1}(\Delta x_i)^{-1}(x_{i+1} - t)], & x_i \leq t \leq x_{i+1}. \end{cases}$$

On $[x_{i-1}, x_i]$, $(r_i)_x(x-t)_+$ is a linear polynomial $p(t)$ such that

$$p(x_i) = 2(\Delta x_i + \Delta x_{i-1}) + \Delta x_{i-1} - 3\Delta x_{i-1}(\Delta x_{i-1})^{-1}\Delta x_i = 2\Delta x_i$$

and

$$p(x_{i-1}) = 2(\Delta x_i + \Delta x_{i-1}) + \Delta x_{i-1} - 3[\Delta x_i(\Delta x_i)^{-1}\Delta x_{i-1} + \Delta x_{i-1}(\Delta x_i)^{-1}(\Delta x_i + \Delta x_{i-1} - \Delta x_{i-1})] = -\Delta x_i.$$

Thus, $p(\frac{2}{3}x_{i-1} + \frac{1}{3}x_i) = 0$. On $[x_i, x_{i+1}]$, $r_i(x - t)_+$ is a linear polynomial $q(t)$ such that

$$q(x_i) = \Delta x_{i-1} - 3\Delta x_{i-1}(\Delta x_i)^{-1}\Delta x_i = -2\Delta x_{i-1}$$

and

$$q(x_{i+1}) = \Delta x_{i-1}.$$

Thus, $q(\frac{1}{3}x_i + \frac{2}{3}x_{i+1}) = 0$, and the graph of $(r_i)_x(x - t)_+$ is as follows:

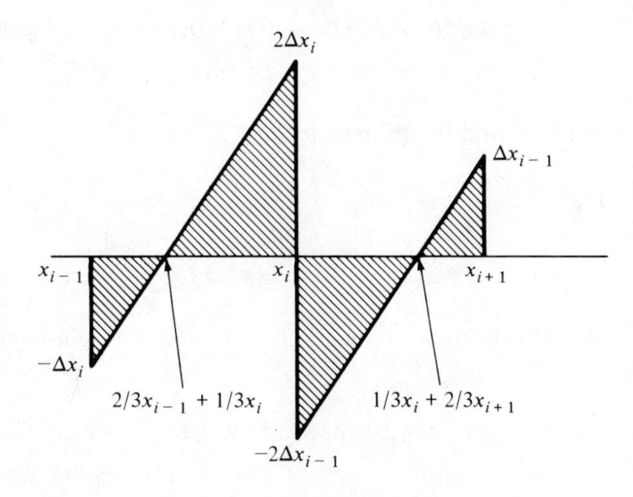

Thus,

$$|r_i(f)| \leq \|D^2f\|_\infty \times \text{(sum of the areas of the shaded triangles of the preceding graph)}$$

$$= \tfrac{1}{2}[\tfrac{1}{3}\Delta x_{i-1}(\Delta x_i) + \tfrac{2}{3}\Delta x_{i-1}(2\Delta x_i) + \tfrac{2}{3}\Delta x_i(2\Delta x_{i-1}) + \tfrac{1}{3}\Delta x_i(\Delta x_{i-1})]\,\|D^2f\|_\infty$$

$$= \tfrac{5}{3}\Delta x_{i-1}\Delta x_i\,\|D^2f\|_\infty.$$

Moreover,

$$|D_{ii}r_i(f)| \leq \tfrac{1}{2}\cdot\tfrac{5}{3}\Delta x_{i-1}\Delta x_i(\Delta x_{i-1} + \Delta x_i)^{-1}\,\|D^2f\|_\infty \leq \tfrac{5}{6}h\,\|D^2f\|_\infty.$$

Since $\|(DB)^{-1}\|_\infty \leq 2$, we have $\|\mathbf{e}^1\|_\infty \leq \frac{5}{3}h\|D^2f\|$. Combining this bound with the results of Lemma 3.1 and Theorem 3.7, we have

$$\|f - \vartheta_S f\|_\infty \leq \tfrac{1}{4}h^2 \|D^2 f\|_\infty + (\tfrac{1}{4}h)(\tfrac{5}{3}h \|D^2 f\|_\infty) = \tfrac{2}{3}h^2 \|D^2 f\|_\infty.$$

Q.E.D.

We now proceed to the a priori error bounds for the bicubic spline interpolation procedure. We find as in the one-dimensional case that if f is sufficiently smooth, then $\vartheta_{S(\rho)} f$ is a fourth-order approximation to f with respect to both the L^∞-norm and the L^2-norm.

THEOREM 4.9

If $f \in PC^{4,2}(U)$, then

$$\begin{aligned} \|f - \vartheta_S f\|_2 &\leq 4\pi^{-4}(h^4 \|D_x^4 f\|_2 + h^2k^2 \|D_x^2 D_y^2 f\|_2 + k^4 \|D_y^4 f\|_2) \\ &\leq 4\pi^{-4}\bar{\rho}^4(\|D_x^4 f\|_2 + \|D_x^2 D_y^2 f\|_2 + \|D_y^4 f\|_2). \end{aligned} \tag{4.42}$$

Proof. From (4.8) and the triangle inequality, we have

$$\begin{aligned} \|f - \vartheta_S f\|_2 &\leq \|f - \vartheta_{S(\Delta)} f\|_2 + \|\vartheta_{S(\Delta)}(f - \vartheta_{S(\Delta_y)} f)\|_2 \\ &\leq \|f - \vartheta_{S(\Delta)} f\|_2 + \|\vartheta_{S(\Delta)}(f - \vartheta_{S(\Delta_y)}) \\ &\qquad - (f - \vartheta_{S(\Delta_y)})\|_2 + \|f - \vartheta_{S(\Delta_y)} f\|_2. \end{aligned} \tag{4.43}$$

Using the results of Theorems 4.5 and 4.6 to bound the right-hand side of (4.43), we have

$$\begin{aligned} \|f - \vartheta_S f\|_2 \leq 4\pi^{-4}h^4 \|D_x^4 f\|_2 &+ 2\pi^{-2}h^2 \|D_x^2(f - \vartheta_{S(\Delta_y)} f\|_2 \\ &+ 4\pi^{-4}k^4 \|D_y^4 f\|_2. \end{aligned} \tag{4.44}$$

But since $D_x^2 \vartheta_{S(\Delta_y)} f = \vartheta_{S(\Delta_y)} D_x^2 f$, we have

$$\|D_x^2(f - \vartheta_{S(\Delta_y)} f)\|_2 \leq 2\pi^{-2}k^2 \|D_x^2 D_y^2 f\|_2. \tag{4.45}$$

Using (4.45) to bound the right-hand side of (4.44), we obtain (4.42). Q.E.D.

Using the results of Theorems 4.7 and 4.8, we may prove the following result in essentially the same way.

THEOREM 4.10

If $f \in PC^{4,\infty}(U)$, then

$$\begin{aligned} \|f - \vartheta_S f\|_\infty &\leq \tfrac{5}{384}h^4 \|D_x^4 f\|_\infty + \tfrac{4}{9}h^2k^2 \|D_x^2 D_y^2 f\|_\infty + \tfrac{5}{384}k^4 \|D_y^4 f\|_\infty \\ &\leq \bar{\rho}^4(\tfrac{5}{384} \|D_x^4 f\|_\infty + \tfrac{4}{9} \|D_x^2 D_y^2 f\|_\infty + \tfrac{5}{384} \|D_y^4 f\|_\infty). \end{aligned} \tag{4.46}$$

EXERCISES FOR CHAPTER 4

(4.1) Using the Peano Kernel Theorem, show that if $f \in PC^{4,\infty}(I)$, then

$$\| D(f - \vartheta_S f) \|_\infty \leq \left(\frac{\sqrt{3}}{216} + \frac{1}{24}\right) h^3 \| D^4 f \|_\infty$$

(cf. [4.6]).

(4.2) Using the Peano Kernel Theorem, show that if $f \in PC^{4,\infty}(I)$, then

$$\| D^2(f - \vartheta_S f) \|_\infty \leq (\tfrac{1}{12} + \tfrac{1}{3}\underline{h}^{-1} h) h^2 \| D^4 f \|_\infty$$

(cf. [4.6]).

(4.3) Using the Peano Kernel Theorem, show that if $f \in PC^{4,\infty}(I)$, then

$$\| D^3(f - \vartheta_S f) \|_\infty \leq \tfrac{1}{2}(1 + \underline{h}^{-2} h^2) h \| D^4 f \|_\infty$$

(cf. [4.6]).

(4.4) Given $\mathbf{f} \equiv (f_0, f_1, \ldots, f_{N+1})$, let $\vartheta^{\text{II}}_{S(\Delta)}\mathbf{f}$, the Type II $S(\Delta)$-interpolate of $\mathbf{f}$, be the unique spline, $s(x)$, in $S(\Delta)$ such that $s(x_i) = f_i$, $0 \leq i \leq N + 1$, and $D^2 s(x_i) = 0$, $i = 0$ and $N + 1$. Show that the mapping $\vartheta^{\text{II}}_{S(\Delta)}$ is well-defined (cf. [4.8], [4.10], and [4.13]).

(4.5) Given $\mathbf{f} \equiv (f_0, \ldots, f_{N+1}, f^2_0, f^2_{N+1})$, let $\vartheta^{\text{III}}_{S(\Delta)}\mathbf{f}$, the Type III, $S(\Delta)$-interpolate of $\mathbf{f}$, be the unique spline, $s(x)$, in $S(\Delta)$ such that $s(x_i) = f_i$, $0 \leq i \leq N + 1$, and $D^2 s(x_i) = f^2_i$, $i = 0$ and $N + 1$. Show that the mapping $\vartheta^{\text{III}}_{S(\Delta)}$ is well-defined (cf. [4.8], [4.10], and [4.13]).

(4.6) Given $f(x) \in C^3_p(I)$, i.e., $f(x) \in C^3(I)$ and $D^k f(0) = D^k f(1) = 0, 0 \leq k \leq 3$, let $\vartheta^{\text{IV}}_{S(\Delta)}\mathbf{f}$, the Type IV, $S(\Delta)$-interpolate of $\mathbf{f}$, be the unique spline, $s(x)$, in $S(\Delta)$ such that $s(x_i) = f_i, 0 \leq i \leq N + 1$, and $D^k s(0) = D^k s(1), 1 \leq k \leq 2$. Show that the mapping $\vartheta^{\text{IV}}_{S(\Delta)}$ is well-defined (cf. [4.8], [4.10], and [4.13]).

(4.7) Show that if $s(x)$ is the Type II or IV interpolate of $f \in PC^{2,2}(I)$, then it satisfies the First Integral Relation (cf. [4.8] and [4.10]).

(4.8) Show that if $s(x)$ is the Type III or IV interpolate of $f \in PC^{4,2}(I)$, then it satisfies the Second Integral Relation (cf. [4.8] and [4.10]).

(4.9) Derive error bounds analogous to those of Section 4.1 for the interpolation procedures of Types II, III, and IV (cf. [4.8], [4.10], and [4.13]).

(4.10) Develop a theory for two-dimensional spline interpolation procedures of Types II, III, and IV, analogous to that of Section 4.2.

(4.11) Use Exercises (1.2) and (1.4) to show that if $f \in PC^{2,2}(I)$, then

$$\| f - \vartheta_S f \|_p \leq \pi^{-1} h^{3/2 + p^{-1}} \| D^2 f \|_2$$

for all $p \geq 2$. Furthermore, show that if $f \in PC^{4,2}(I)$, then

$$\| f - \vartheta_S f \|_p \leq 2\pi^{-3} h^{7/2 + p^{-1}} \| D^4 f \|_2$$

for all $p \geq 2$.

(4.12) If m is a positive integer and z is an integer such that $m - 1 \leq z \leq 2m - 2$, we define the *spline space* $S(2m - 1, \Delta, z)$ to be the set of all real-valued functions $s(x) \in C^z(I)$ such that on each subinterval $[x_i, x_{i+1}]$, $0 \leq i \leq N$, $s(x)$ is a polynomial of degree $2m - 1$. Moreover, we define the interpolation mapping $\vartheta_m: C^{m-1}(I) \longrightarrow S(2m - 1, \Delta, z)$ by $\vartheta_m f \equiv s$, where

$$D^k s(x_i) = D^k f(x_i), \begin{cases} 0 \leq k \leq 2m - 2 - z, & 1 \leq i \leq N, \\ 0 \leq k \leq m - 1, & i = 0 \text{ and } N + 1. \end{cases}$$

Show that $S(1, \Delta, 0) = L(\Delta)$, $S(3, \Delta, 1) = H(\Delta)$, and $S(3, \Delta, 2) = S(\Delta)$. Moreover, show that the interpolation mapping ϑ_m is well-defined (cf. [4.8], [4.10], and [4.13]).

(4.13) Using the notations of Exercise (4.12), show that if $f \in PC^{m,2}(I)$, then $\vartheta_m f$ satisfies the First Integral Relation, i.e.,

$$\| D^m f \|_2^2 = \| D^m(f - \vartheta_m f)\|_2^2 + \| D^m \vartheta_m f \|_2^2$$

(cf. [4.8], [4.10], and [4.13]).

(4.14) Using the notations of Exercise (4.12), show that if $f \in PC^{2m,2}(I)$, then $\vartheta_m f$ satisfies the Second Integral Relation, i.e.,

$$\| D^m(f - \vartheta_m f)\|_2^2 = (f - \vartheta_m f, D^{2m} f)_2.$$

(4.15) Using the notations of Exercise (4.12), show that if $f \in PC^{m,2}(I)$, then

$$\| D^j(f - \vartheta_m f)\|_2 \leq K_{m,m,z,j} h^{m-j} \| D^m f \|_2, \qquad 0 \leq j \leq m,$$

where

$$K_{m,m,z,j} \equiv \begin{cases} 1, & m - 1 \leq z \leq 2m - 2, \ j = m, \\ \pi^{-m+j}, & m - 1 = z, \ 0 \leq j \leq m - 1, \\ \pi^{-m+j}(z + 2 - m)!, & m - 1 \leq z \leq 2m - 2, \\ & \qquad 0 \leq j \leq 2m - 2 - z, \\ \pi^{-m+j}(z + 2 - \mathrm{m})!(j!)^{-1}, & m - 1 \leq z \leq 2m - 2, \\ & \qquad 2m - 2 - z \leq j \leq m - 1. \end{cases}$$

Moreover, show that if $f \in PC^{2m,2}(I)$, then

$$\| D^j(f - \vartheta_m f)\|_2 \leq K_{m,2m,z,j} h^{2m-j} \| D^{2m} f \|_2, \qquad 0 \leq j \leq m,$$

where $K_{m,2m,z,j} \equiv (K_{m,m,z,j})(K_{m,m,z,0})$ (cf. [4.11]).

(4.16) Using the notations of Exercise (4.12), let ϑ_m^* be the product interpolation mapping into the tensor product space $S(2m - 1, \Delta, z) \otimes S(2m - 1, \Delta_y, z)$ and show that if $f \in PC^{2m,2}(R)$, then

$$\begin{aligned} \|f - \vartheta_m^* f\|_2 &\leq K_{m,2m,z,0}(h^{2m} \| D_x^{2m} f\|_2 + h^m k^m \| D_x^m D_y^m f \|_2 + k^{2m} \| D_y^{2m} f \|_2) \\ &\leq K_{m,2m,z,0} \bar{\rho}^{2m}(\| D_x^{2m} f \|_2 + \| D_x^m D_y^m f \|_2 + \| D_y^{2m} f \|_2). \end{aligned}$$

(4.17) Using the notations of Exercise (4.12) and the results of Exercises (1.2) and (1.4), show that if $f \in PC^{m,2}(I)$, then

$$\|f - \vartheta_m f\|_p \leq \tfrac{1}{2} K_{m,m,z,1} h^{m-1/2+p^{-1}} \|D^m f\|_2, \qquad p \geq 2.$$

Moreover, show that if $f \in PC^{2m,2}(I)$, then

$$\|f - \vartheta_m f\|_p \leq \tfrac{1}{2} K_{m,2m,z,1} h^{2m-1/2+p^{-1}} \|D^{2m} f\|_2, \qquad p \geq 2.$$

(4.18) Develop analogues of the results of Exercises (4.12)–(4.17) for interpolation mappings of Types II, III, and IV into $S(2m - 1, \Delta, z)$ (cf. [4.8], [4.10], and [4.13]).

REFERENCES FOR CHAPTER 4

[4.1] AHLBERG, J. H., E. N. NILSON, and J. L. WALSH, Convergence properties of generalized splines. *Proc. Nat. Acad. Sci. U.S.A.* **54**, 344–350 (1965).

[4.2] BIRKHOFF, G., and C. DEBOOR, Error bounds for spline interpolation. *J. Math. Mech.* **13**, 827–836 (1964).

[4.3] CARLSON, R. E., and C. A. HALL, On piecewise polynomial interpolation in rectangular polygons. *J. Approx. Theory* **4**, 37–53 (1971).

[4.4] COURANT, R., and D. HILBERT, *Methods of Mathematical Physics*. Interscience, New York (1962).

[4.5] DEBOOR, C., Bicubic spline interpolation. *J. Math. Phys.* **41**, 212–218 (1962).

[4.6] HALL, C. A., On error bounds for spline interpolation. *J. Approx. Theory* **1**, 209–218 (1968).

[4.7] ISAACSON, E., and H. B. KELLER, *Analysis of Numerical Methods*. John Wiley & Sons, Inc., New York (1966).

[4.8] LUCAS, T. R., A generalization of *L*-splines. *Numer. Math.* **15**, 359–370 (1970).

[4.9] SCHOENBERG, I. J., Contributions to the problem of approximation of equidistant data by analytic functions. Parts A and B. *Quart. Appl. Math.* **4**, 45–99, 112–141 (1946).

[4.10] SCHULTZ, M. H., and R. S. VARGA, *L*-splines. *Numer. Math.* **10**, 345–369 (1967).

[4.11] SCHULTZ, M. H., Error bounds for polynomial spline interpolation. *Math. of Comp.* **24**, 507–515. (1970).

[4.12] SCHULTZ, M. H., *Computing spline interpolates without derivatives.* Yale Computer Science Research Report.

[4.13] SWARTZ, K. B., and R. S. Varga, Bounds for some spline interpolation errors. (to appear).

5 LINEAR INTEGRAL EQUATIONS

In this chapter we discuss the method of degenerate kernels for approximating the solutions of Fredholm integral equations of Type II. In particular, we consider the use of bivariate, piecewise polynomial kernels.

We consider the equation

$$u(x) = \int_0^1 K(x, y)u(y)\,dy + f(x), \qquad 0 \le x \le 1, \tag{5.1}$$

where $K(x, y)$ and $f(x)$ are given real-valued continuous functions. The kernel $K(x, y)$ is said to be *degenerate* if it has the form

$$K(x, y) = \sum_{l=0}^{n} \sum_{k=0}^{m} \beta_{lk} B_l(x) C_k(y). \tag{5.2}$$

Substituting (5.2) into (5.1), we see that if $u(x)$ is a solution, then

$$u(x) = \sum_{l=0}^{n} B_l(x) \int_0^1 \sum_{k=0}^{m} \beta_{lk} C_k(y) u(y)\,dy + f(x) \tag{5.3}$$

or

$$u(x) = f(x) + \sum_{l=0}^{n} \alpha_l B_l(x) \tag{5.4}$$

where

$$\alpha_l \equiv \int_0^1 \sum_{k=0}^{m} \beta_{lk} C_k(y) u(y)\,dy. \tag{5.5}$$

Equation (5.4) determines the structure of $u(x)$, and it remains only to determine the coefficient vector $\boldsymbol{\alpha}$. Substituting (5.4) into (5.1), we get

$$(5.6)\qquad \sum_{i=0}^{n} \alpha_i B_i(x) = \int_0^1 \sum_{l=0}^{n} \sum_{k=0}^{m} \beta_{lk} B_l(x) C_k(y) \Big(\sum_{j=0}^{n} \alpha_j B_j(y) + f(y) \Big)\, dy.$$

Setting the coefficient of $B_i(x)$ on the left-hand side of (5.6) equal to the coefficient of $B_i(x)$ on the right-hand side of (5.6) for each $0 \le i \le n$, we have

$$(5.7)\qquad \alpha_i = \sum_{k=0}^{m} \int_0^1 \beta_{ik} C_k(y) \sum_{j=0}^{n} \alpha_j B_j(y)\, dy + \sum_{k=0}^{m} \int_0^1 \beta_{ik} C_k(y) f(y)\, dy$$

for $0 \le i \le n$. Putting this linear system into matrix form, we have

$$(5.8)\qquad A\boldsymbol{\alpha} = \mathbf{k},$$

where $A = I - B$, $B \equiv [b_{ij}]$,

$$b_{ij} \equiv \sum_{k=0}^{m} \int_0^1 \beta_{ik} C_k(y) B_j(y)\, dy, \qquad 0 \le i, j \le n,$$

$$\mathbf{k} \equiv [k_i],$$

and

$$k_i \equiv \sum_{k=0}^{m} \int_0^1 \beta_{ik} C_k(y) f(y)\, dy, \qquad 0 \le i \le n.$$

Of course in general, $K(x, y)$ is not degenerate, but we can approximate it by a piecewise bivariate polynomial, $P(x, y)$, which is degenerate. In this chapter, we will concentrate on approximating the kernel by interpolation. However, it is clear that we could as well approximate it by a least squares technique. Moreover, in a problem in which $K(x, y)$ is experimentally determined, this latter technique might be preferable.

We first study the question of the nonsingularity of the matrix A for a degenerate kernel $P(x, y)$ approximating $K(x, y)$. If we rewrite (5.1) in operator form as

$$(5.9)\qquad (I - K)u = f$$

and the approximate integral equation (which reduces to (5.8)) as

$$(5.10)\qquad (I - P)\bar{u} = f,$$

then we show that if $\|K - P\|_\infty \equiv \max_{0 \le x, y \le 1} |K(x, y) - P(x, y)|$ is sufficiently small and if $I - K$ is invertible (i.e., $(I - K)u = f$ has a unique solution $u(x) \in C[0, 1]$ for all $f \in C[0, 1]$), and if there exists a positive constant, which we will denote by $\|(I - K)^{-1}\|_\infty$, independent of f such that $\|u\|_\infty \le \|(I - K)^{-1}\|_\infty \|f\|_\infty$ for all $f \in C[0, 1]$, then $I - P$ is invertible.

THEOREM 5.1

If $I - K$ is invertible and $\|K - P\|_\infty \|(I - K)^{-1}\|_\infty = q < 1$, then $I - P$ is invertible.

Proof. We must show that for all $f \in C[0, 1]$, $(I - P)u = f$ has a unique solution in $C[0, 1]$. Let $g \in C[0, 1]$ be such that $(I - K)g = f$ and $\|g\|_\infty \leq \|(I - K)^{-1}\|_\infty \|f\|_\infty$. Then it suffices to consider $(I - K)^{-1}(I - P)u = g$ and to show that this has a unique solution. But

$$\begin{aligned}(I - K)^{-1}(I - P) &= I - (I - K)^{-1}[(I - K) - (I - P)] \\ &= I - (I - K)^{-1}(P - K),\end{aligned}$$

and

$$\|(I - K)^{-1}(P - K)\|_\infty \leq \|(I - K)^{-1}\|_\infty \|P - K\|_\infty = q < 1.$$

Thus if $W \equiv (I - K)^{-1}(P - K)$, we want to solve $(I - W)u = g$ or $u = Wu + g \equiv V(u)$ where V satisfies

$$\|V(u) - V(z)\|_\infty = \|W(u - z)\|_\infty \leq q\|u - z\|_\infty$$

and is a contraction mapping on $C[0, 1]$. By the contraction mapping theorem (cf. [5.1]), $(I - K)^{-1}(I - P)$ is invertible and

$$\|u\|_\infty \leq (1 - q)^{-1}\|(I - K)^{-1}\|_\infty + \|f\|_\infty.$$

Q.E.D.

We can now obtain two general error bounds. The first is an a priori bound and depends on $\|u\|_\infty$ and the second is an a posteriori bound and depends on $\|\bar{u}\|_\infty$.

THEOREM 5.2

If the hypotheses of Theorem 5.1 hold,

$$\|u - \bar{u}\|_\infty \leq q(1 - q)^{-1}\|u\|_\infty, \tag{5.11}$$

and

$$\|u - \bar{u}\|_\infty \leq q\|\bar{u}\|_\infty. \tag{5.12}$$

Proof.

$$\begin{aligned}\bar{u} &= (I - P)^{-1}f = (I - P)^{-1}(I - K)u \\ &= (I - P)^{-1}[(I - P) + (P - K)]u \\ &= u + (I - P)^{-1}(P - K)u\end{aligned}$$

and hence

$$\begin{aligned} \|u - \bar{u}\|_\infty &\leq \|(I - P)^{-1}(P - K)\|_\infty \|u\|_\infty \\ &\leq \frac{\|(I - K)^{-1}\|_\infty \|P - K\|_\infty}{1 - \|(I - K)^{-1}\|_\infty \|P - K\|_\infty} \|u\|_\infty, \end{aligned}$$

which proves (5.11). Likewise,

$$\begin{aligned} u &= (I - K)^{-1}f = (I - K)^{-1}(I - P)\bar{u} \\ &= (I - K)^{-1}[(I - K) + (K - P)]\bar{u} = \bar{u} + (I - K)^{-1}(K - P)\bar{u}, \end{aligned}$$

and hence

$$\|u - \bar{u}\|_\infty \leq \|(I - K)^{-1}\|_\infty \|K - P\|_\infty \|\bar{u}\|_\infty.$$

Q.E.D.

From Theorem 2.8, we have the following result.

COROLLARY 1

If $I - K$ is invertible, $K(x, y) \in PC^{2,\infty}(U)$, and ρ is such that

$$q^L_{h,k} \equiv \tfrac{1}{8}(h^2 \|D_x^2 K\|_\infty + k^2 \|D_y^2 K\|_\infty) \|(I - K)^{-1}\|_\infty < 1,$$

then $I - \vartheta_{L(\rho)}K$ is invertible and

$$\|u - \bar{u}_L\|_\infty \leq q^L_{h,k}(1 - q^L_{h,k})^{-1} \|u\|_\infty, \tag{5.13}$$

i.e., we have a second-order approximation scheme.

From Theorems 3.10 and 4.10, we have the following results.

COROLLARY 2

If $I - K$ is invertible, $K(x, y) \in PC^{4,\infty}(U)$ and ρ is such that

$$q^H_{h,k} \equiv \tfrac{1}{384}(h^4 \|D_x^4 K\|_\infty + 24h^2k^2 \|D_x^2 D_y^2 K\|_\infty + k^4 \|D_y^4 K\|_\infty) \|(I - K)^{-1}\|_\infty < 1,$$

then $I - \vartheta_{H(\rho)}K$ is invertible and

$$\|u - \bar{u}_H\|_\infty \leq q^H_{h,k}(1 - q^H_{h,k})^{-1} \|u\|_\infty, \tag{5.14}$$

i.e., we have a fourth-order approximation scheme.

COROLLARY 3

If $I - K$ is invertible, $K(x, y) \in PC^{4,\infty}(U)$ and ρ is such that

$$\begin{aligned} q^S_{h,k} \equiv (\tfrac{5}{384}h^4 \|D_x^4 K\|_\infty + \tfrac{4}{9}h^2k^2 \|D_x^2 D_y^2 K\|_\infty + \tfrac{5}{384}k^4 \|D_y^4 K\|_\infty) \\ \times \|(I - K)^{-1}\|_\infty < 1, \end{aligned}$$

then $I - \vartheta_{S(\rho)}K$ is invertible and

$$\|u - \tilde{u}_S\|_\infty \leq q^S_{h,k}(1 - q^S_{h,k})^{-1}\|u\|_\infty, \tag{5.15}$$

i.e., we have a fourth-order approximation scheme.

EXERCISE FOR CHAPTER 5

(5.1) Using the notations and results of Exercise 3.19, show that if $I - K$ is invertible and $K(x, y) \in PC^{2m,\infty}(R)$, then there exists a positive constant, C, such that if

$$q^m_{h,k} \equiv C(h^{2m}\|D^{2m}_x K\|_\infty + h^m k^m \|D^m_x D^m_y K\|_\infty + k^{2m}\|D^{2m}_y K\|_\infty) \times \|(I - K)^{-1}\|_\infty < 1,$$

then $I - \vartheta_{H^m(\rho)}K$ is invertible and

$$\|u - \tilde{u}_{H^m}\|_\infty \leq q^m_{h,k}(1 - q^m_{h,k})^{-1}\|u\|_\infty,$$

i.e., we have a $2m$th-order approximation scheme.

REFERENCE FOR CHAPTER 5

[5.1] GOFFMAN, C., and G. PEDRICK, *First Course in Functional Analysis*. Prentice-Hall, Inc., Englewood Cliffs, N. J., (1967).

6 FINITE ELEMENT REGRESSION

6.1 ONE-DIMENSIONAL PROBLEMS

In this chapter, we study finite element regression or least squares approximation by means of piecewise polynomial functions. Generally, we recommend least squares procedures for smoothing "noisy" or oscillatory data when we wish to avoid the extraneous oscillations of the approximations given by interpolation procedures. We start by considering general one-dimensional problems.

Let $\{B_i(x)\}_{i=1}^n$ denote n linearly independent basis functions in $PC^{0,2}(I)$ and $f \in PC^{0,2}(I)$. We consider the least squares variational problem of finding $\boldsymbol{\beta}^* \in R^n$ such that

$$\phi(\boldsymbol{\beta}^*) = \inf_{\boldsymbol{\beta} \in R^n} \phi(\boldsymbol{\beta}) \equiv \inf_{\boldsymbol{\beta} \in R^n} \left\| f - \sum_{i=1}^n \beta_i B_i \right\|_2^2. \tag{6.1}$$

The function

$$\phi(\boldsymbol{\beta}) = \|f\|_2^2 - 2(f, \sum_{i=1}^n \beta_i B_i)_2 + \left\| \sum_{i=1}^n \beta_i B_i \right\|_2^2$$

is clearly quadratic in $\boldsymbol{\beta} \in R^n$ and hence $\boldsymbol{\beta}^*$ is a solution of (6.1) if and only if

$$D_i \phi(\boldsymbol{\beta}^*) = 0, \qquad 1 \leq i \leq n, \tag{6.2}$$

and the matrix $J[\boldsymbol{\beta}^*] \equiv [D_i D_j \phi(\boldsymbol{\beta}^*)]$ is positive definite. Carrying out the differentiation in (6.2), we obtain the linear system

$$A\boldsymbol{\beta}^* = \mathbf{k}, \tag{6.3}$$

where $A \equiv [a_{ij}]$, $a_{ij} \equiv (B_i, B_j)_2$, $1 \leq i, j \leq n$, $\mathbf{k} \equiv [k_i]$, and $k_i \equiv (f, B_i)_2$,

$1 \leq i \leq n$. Moreover,

(6.4) $$J[\boldsymbol{\beta}] = 2A, \qquad \text{for all } \boldsymbol{\beta} \in R^n.$$

Using (6.3) and (6.4), we have the following result.

THEOREM 6.1

The least squares variational problem of finding $\boldsymbol{\beta}^*$ satisfying (6.1) has a unique solution,

$$\sum_{i=1}^{N} \beta_i^* B_i(x),$$

where the $\boldsymbol{\beta}^*$ is the solution of a symmetric, positive definite linear system.

Proof. The matrix A is clearly symmetric. If $\boldsymbol{\beta} \neq \mathbf{0}$,

$$\boldsymbol{\beta}^T A \boldsymbol{\beta} = \left\| \sum_{i=1}^{n} \beta_i B_i \right\|_2^2 > 0.$$

In fact, if $\left\| \sum_{i=1}^{n} \beta_i B_i \right\|_2 = 0$, then, since $\{B_i\}_{i=1}^n$ are linearly independent, $\boldsymbol{\beta} = \mathbf{0}$, which contradicts the choice of $\boldsymbol{\beta}$. Hence, A is positive definite.

Furthermore, (6.3) has a unique solution $\boldsymbol{\beta}^*$. Since $J[\boldsymbol{\beta}^*] = 2A$, $\boldsymbol{\beta}^*$ is actually a solution of the variational problem (6.1). Q.E.D.

If $S \equiv \left\{ \sum_{i=1}^{n} \beta_i B_i(x) \,|\, \boldsymbol{\beta} \in R^n \right\}$, then the unique solution given by (6.3) is denoted by $P_S f \equiv \sum_{i=1}^{n} \beta_i^* B_i(x)$. Clearly $P_S f$ is the orthogonal projection of f onto S with respect to the L^2-inner product; cf. [6.4].

We now examine the question of choice of basis functions so as to yield sparse, well-conditioned matrices.

As a basis for $L(\Delta)$, we suggest the functions $\{l_i(x)\}_{i=0}^{N+1}$ defined in Section 2.1. Using these to form the system (6.3), we obtain a system with a tridiagonal matrix. Such systems can be solved very efficiently by Gaussian elimination; cf. [6.2] and [6.6]. Moreover, in the special case of a uniform partition, i.e., $x_i \equiv ih$, $0 \leq i \leq N + 1$, where $h \equiv (N + 1)^{-1}$, the matrix $A_h \equiv [a_{ij}]$ is given by

(6.5) $$A_h \equiv \frac{h}{6} \begin{bmatrix} 2 & 1 & & & \mathbf{0} \\ 1 & 4 & \ddots & & \\ & \ddots & \ddots & \ddots & \\ & & \ddots & 4 & 1 \\ \mathbf{0} & & & 1 & 2 \end{bmatrix},$$

where the solid lines indicate a continuation of the same entry. Since A_h is symmetric, its eigenvalues are real, and by the Gerschgorin Theorem (cf. [6.6] and [6.11]), the eigenvalues lie in the open interval $(h/6, h)$. Thus, the condition number of A_h, cond (A_h), which for symmetric, positive definite matrices is the ratio of the maximum eigenvalue to the minimum eigenvalue, satisfies

$$\text{cond}\,(A_h) < 6. \tag{6.6}$$

For a detailed discussion of the computational significance of condition numbers, see [6.2] and [6.6]. The important thing to note in the bound (6.6) is that the condition number is bounded *independent* of h.

For the case of nonuniform partitions, we can obtain a similar result, though the proof is different.

THEOREM 6.2

If $A_{L(\Delta)}$ is the least squares matrix (6.3) obtained by using the basis functions $\{l_i(x)\}_{i=0}^{N+1}$ defined in Section 2.1, then

$$\text{cond}\,(A_{L(\Delta)}) \leq 6(\underline{h}^{-1}h). \tag{6.7}$$

Proof. Since $A_{L(\Delta)}$ is symmetric and positive definite, it suffices to find positive numbers λ and Λ such that

$$\lambda\Big(\sum_{i=0}^{N+1} \beta_i^2\Big) \leq \boldsymbol{\beta}^T A_{L(\Delta)} \boldsymbol{\beta} = \int_0^1 \Big[\sum_{i=0}^{N+1} \beta_i l_i(x)\Big]^2 dx \leq \Lambda\Big(\sum_{i=0}^{N+1} \beta_i^2\Big).$$

In fact, then cond $(A_{L(\Delta)}) \leq \lambda^{-1}\Lambda$. If we let

$$\begin{aligned} t_j &\equiv \int_{x_j}^{x_{j+1}} \Big[\sum_{i=0}^{N+1} \beta_i l_i(x)\Big]^2 dx \\ &= \int_{x_j}^{x_{j+1}} [\beta_j l_j(x) + \beta_{j+1} l_{j+1}(x)]^2\, dx, \qquad 0 \leq j \leq N, \end{aligned}$$

then by a change of variables we have

$$\begin{aligned} t_j &= (x_{j+1} - x_j) \int_0^1 [\beta_j \tilde{l}_0(y) + \beta_{j+1} \tilde{l}_1(y)]^2\, dy \\ &\equiv (x_{j+1} - x_j) I_j, \end{aligned} \tag{6.8}$$

where $\tilde{l}_0(y) \equiv 1 - y$ and $\tilde{l}_1(y) \equiv y$ are *independent* of j, for all $0 \leq j \leq N$. But the integral $I_j = [\beta_j, \beta_{j+1}] M [\beta_j, \beta_{j+1}]^T$, where M is the 2×2 matrix

$$\begin{bmatrix} \frac{1}{3} & \frac{1}{6} \\ \frac{1}{6} & \frac{1}{3} \end{bmatrix}, \tag{6.9}$$

whose eigenvalues are $\frac{1}{6}$ and $\frac{3}{6}$. Thus,

$$\frac{1}{6}(\beta_j^2 + \beta_{j+1}^2)\underline{h} \leq t_j \leq \frac{3}{6}(\beta_j^2 + \beta_{j+1}^2)h, \qquad 0 \leq j \leq N. \tag{6.10}$$

Summing the inequality (6.10) with respect to j from 0 to N, we obtain

$$\begin{aligned} \tfrac{1}{6}\underline{h}\Big(\sum_{i=0}^{N+1} \beta_i^2\Big) &\leq \tfrac{1}{6}\underline{h}\Big(\beta_0^2 + 2\sum_{i=1}^{N} \beta_i^2 + \beta_{N+1}^2\Big) \\ &\leq \boldsymbol{\beta}^T A_{L(\Delta)} \boldsymbol{\beta} \\ &\leq \tfrac{3}{6}h\Big(\beta_0^2 + 2\sum_{i=1}^{N} \beta_i^2 + \beta_{N+1}^2\Big) \\ &\leq h\Big(\sum_{i=0}^{N+1} \beta_i^2\Big). \end{aligned} \tag{6.11}$$

Hence, we may take $\lambda \equiv \frac{1}{6}\underline{h}$ and $\Lambda \equiv h$. Q.E.D.

As a basis for $H(\Delta)$, we suggest the functions $\{h_i(x)\}_{i=0}^{N+1}$ and

$$\tilde{h}_i^1(x) \equiv \begin{cases} \frac{1}{2}x_1^{-1}h_0^1(x), & i = 0, \\ (x_{i+1} - x_{i-1})^{-1}h_i^1(x), & 1 \leq i \leq N, \\ \frac{1}{2}(1 - x_N)^{-1}h_{N+1}^1(x), & i = N+1, \end{cases} \tag{6.12}$$

where the functions $h_i(x)$ and $h_i^1(x)$ are defined in Section 3.1 and we have normalized the functions $h_i^1(x)$ so as to make all the diagonal entries in $A_{H(\Delta)}$ the same order of magnitude. It is easily verified that $A_{H(\Delta)}$ is six-diagonal and that if we couple together $h_i(x)$ and $\tilde{h}_i^1(x)$ for each $0 \leq i \leq N+1$, then the corresponding (block) matrix is block tridiagonal with 2×2 subblocks.

For the case of uniform partitions of I, we can prove an analogue of Theorem 6.2 for these basis functions; cf. [6.3].

THEOREM 6.3

If Δ is a uniform partition and $A_{H(\Delta)}$ is the least squares matrix (6.3) obtained by using the basis functions $\{h_i(x)/\|h_i(x)\|_2, \tilde{h}_i^1(x)/\|\tilde{h}_i^1(x)\|_2\}_{i=0}^{N+1}$, then

$$\text{cond } A_{H(\Delta)} \leq 160. \tag{6.13}$$

The important thing to note is that the bound of (6.13) is *independent* of h. Moreover, we can ask whether or not there exist other "obvious" basis functions for $H(\Delta)$, which would yield an even better result than (6.13). To this end, for each $\alpha > 0$, we let

$$b_i^\alpha(x) \equiv h_i(x) + \alpha\tilde{h}_i^1(x), \qquad 0 \leq i \leq N+1, \tag{6.14}$$

and

$$(6.15)\qquad t_i^\alpha(x) \equiv h_i(x) - \alpha \tilde{h}_i^1(x), \qquad 0 \leq i \leq N+1,$$

be a basis for $H(\Delta)$. If $A_{H(\Delta)}(\alpha)$ denotes the least squares matrix with respect to these basis functions, then we can seek $\alpha^* > 0$ which minimizes cond $A_{H(\Delta)}(\alpha)$. As a step in this direction, it has been shown that cond $A_{H(\Delta)}(15.5) \approx 26$; cf. [6.3].

To define a basis for $S(\Delta)$, we augment the partition Δ to form $\tilde{\Delta}$: $-x_{-3} < x_{-2} < \ldots < x_{N+1} < \ldots < x_{N+4}$, where, for example, we choose $x_{i+1} - x_i \equiv x_1 - x_0$, $-3 \leq i \leq -1$, and $x_{j+1} - x_j \equiv x_{N+1} - x_N$, $N+1 \leq j \leq N+3$. Following [6.8], we suggest as a basis for $S(\Delta)$ the following "B-splines:"

$$(6.16)\qquad s_i(x) \equiv \sum_{k=0}^{4} [D\omega_i(x_{i+k})]^{-1}(x_{i+k} - x)_+^3, \qquad -3 \leq i \leq N,$$

where $\omega_i(x) \equiv \prod_{k=0}^{4} (x - x_{i+k})$ and

$$y_+^3 \equiv \begin{cases} y^3, & y \geq 0, \\ 0, & y < 0. \end{cases}$$

The graph of $s_i(x)$ is given by

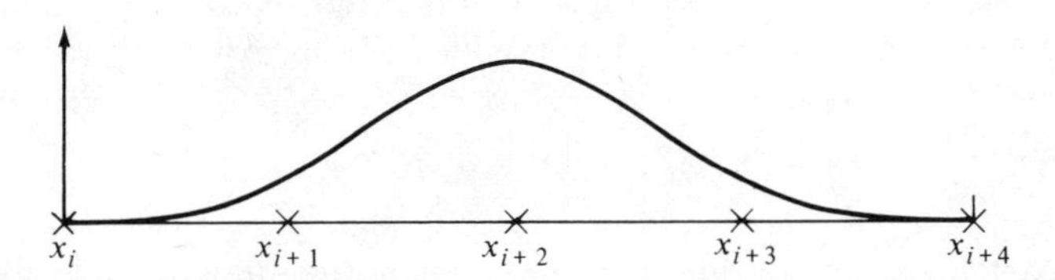

In the special case of a uniform partition with mesh length $h = (N+1)^{-1}$, the basis functions $s_i(x)$, $-3 \leq i \leq N$, can be expressed in terms of a "standard" basis function, $S(x)$. In fact, if

$$S(x) \equiv \begin{cases} (2-x)^3/24 - (1-x)^3/6 - x^3/4 + (1+x)^3/6, & -2 \leq x \leq -1, \\ (2-x)^3/24 - (1-x)^3/6 - x^3/4, & -1 \leq x \leq 0 \\ (2-x)^3/24 - (1-x)^3/6, & 0 \leq x \leq 1 \\ (2-x)^3/24, & 1 \leq x \leq 2 \\ 0, & x \in R - [-2, 2], \end{cases}$$

then $s_i(x) = S(h^{-1}x - i - 2)$, $-3 \leq i \leq N$.

It is easily verified that with these basis functions, $A_{S(\Delta)}$ is a band matrix with seven nonzero diagonals, and hence the linear system (6.3) can be efficiently solved by Gaussian elimination for band matrices. Moreover, for the case of uniform partitions of I, we can prove an analogue of Theorems 6.2 and 6.3 for these basis functions; cf. [6.9].

Now we discuss the question of what to do if the data is given only at the points $T \equiv \{t_i\}_{i=1}^Q$. Our first approach to this problem follows [6.7]. We note that the only place in the linear system (6.3) where the data $f(x)$ plays a role is in the formation of the right-hand side $\mathbf{k} \equiv \left[\int_0^1 f(x)B_i(x)\,dx\right]$.

The idea is to approximate these integrals by expanding $f(x)$ in terms of the data, i.e.,

$$f(x) \approx \sum_{i=1}^{Q} f(t_i)v_i(x) \equiv \tilde{f}(x),$$

in such a way that the integrals $\int_0^1 \tilde{f}(x)B_j(x)\,dx$, $1 \le j \le n$, can be evaluated analytically and such that the error introduced by this procedure is "asymptotically consistent" with the error of the basic least squares method. For the special case of $L(\Delta)$, we suggest letting $\tilde{f} \equiv \vartheta_{L(T)}f$, i.e., $\tilde{f}$ is the piecewise linear interpolate of $f(x)$ with respect to the partition T. For the special cases of $H(\Delta)$ and $S(\Delta)$, we can do essentially the same thing. However, here we suggest using piecewise, cubic Lagrange interpolation with respect to the partition T to form $\tilde{f}$. This procedure was described in detail in Chapters 2 and 3 in the context of approximating derivatives of $f(x)$.

Our second approach to the problem of what to do if $f(x)$ is given only at the points $T \equiv \{t_i\}_{i=1}^Q$ is based on the idea of approximating the functional

$$\phi(\boldsymbol{\beta}) \equiv \left\| f - \sum_{i=1}^{n} \beta_i B_i \right\|_2^2.$$

In particular, we consider the approximate functional

$$\hat{\phi}(\boldsymbol{\beta}) \equiv \sum_{j=1}^{Q} w_j \left[f(t_j) - \sum_{i=1}^{n} \beta_i B_i(t_j) \right]^2,$$

where

$$w_j \equiv \begin{cases} (t_{j+1} - t_j), & j = 1, \\ (t_{j+1} - t_{j-1}), & 2 \le j \le Q - 1, \\ (t_j - t_{j-1}), & j = Q, \end{cases}$$

and define $\hat{\boldsymbol{\beta}}^*$ as the solution of the approximate variational problem of finding $\hat{\boldsymbol{\beta}}^* \in R^n$ such that

$$\hat{\phi}(\hat{\boldsymbol{\beta}}^*) = \inf_{\hat{\boldsymbol{\beta}} \in R^n} \hat{\phi}(\hat{\boldsymbol{\beta}}).$$

It is possible to analyze this procedure completely; cf. [6.9] for the details.

We end this section with some numerical results due to Patent (cf. [6.7]) concerning the approximation of the exponential function, e^x, by finite element regression with the spaces $L(\Delta)$, $H(\Delta)$, and $S(\Delta)$. In all of the following examples the partitions $\Delta(h)$ will be uniform with mesh length h.

$L(\Delta(h))$

h	$\|e^x - P_{L(\Delta(h))}e^x\|_2$	$\|e^x - P_{L(\Delta(h))}e^x\|_\infty$
1	$.63 \times 10^{-1}$	$.16 \times 10^{0}$
2^{-1}	$.17 \times 10^{-1}$	$.51 \times 10^{-1}$
2^{-2}	$.42 \times 10^{-2}$	$.14 \times 10^{-1}$
2^{-3}	$.11 \times 10^{-2}$	$.35 \times 10^{-2}$
2^{-4}	$.27 \times 10^{-3}$	$.88 \times 10^{-3}$
2^{-5}	$.66 \times 10^{-4}$	$.22 \times 10^{-3}$

$H(\Delta(h))$

h	$\|e^x - P_{H(\Delta(h))}e^x\|_2$	$\|e^x - P_{H(\Delta(h))}e^x\|_\infty$
1	$.34 \times 10^{-3}$	$.11 \times 10^{-2}$
2^{-1}	$.43 \times 10^{-4}$	$.15 \times 10^{-3}$
2^{-2}	$.44 \times 10^{-5}$	$.14 \times 10^{-4}$
2^{-3}	$.34 \times 10^{-6}$	$.93 \times 10^{-6}$
2^{-4}	$.23 \times 10^{-7}$	$.62 \times 10^{-7}$
2^{-5}	$.16 \times 10^{-8}$	$.60 \times 10^{-8}$

$S(\Delta(h))$

h	$\|e^x - P_{S(\Delta(h))}e^x\|_2$	$\|e^x - P_{S(\Delta(h))}e^x\|_\infty$
1	$.34 \times 10^{-3}$	$.11 \times 10^{-2}$
2^{-1}	$.46 \times 10^{-4}$	$.19 \times 10^{-3}$
2^{-2}	$.54 \times 10^{-5}$	$.11 \times 10^{-4}$
2^{-3}	$.37 \times 10^{-6}$	$.81 \times 10^{-6}$
2^{-4}	$.24 \times 10^{-7}$	$.56 \times 10^{-7}$
2^{-5}	$.16 \times 10^{-8}$	$.40 \times 10^{-8}$

These numerical results indicate that the piecewise linear least squares approximation to e^x is second-order accurate in both the L^2-norm and the L^∞-norm, while the piecewise cubic Hermite and spline least squares approximations to e^x are fourth-order accurate in both the L^2-norm and the L^∞-norm. In Section 6.3, we will prove that these special results are true for all sufficiently smooth functions.

6.2 TWO-DIMENSIONAL PROBLEMS

Let $\{B_i(x, y)\}_{i=1}^{n}$ denote n linearly independent basis functions in $PC^{0,2}(U)$ and $f \in PC^{0,2}(U)$. We consider the least squares variational problem of finding $\boldsymbol{\beta}^* \in R^n$ such that

$$\phi(\boldsymbol{\beta}^*) = \inf_{\boldsymbol{\beta} \in R^n} \phi(\boldsymbol{\beta}) \equiv \inf_{\boldsymbol{\beta} \in R^n} \int_0^1 \int_0^1 \left[f(x, y) - \sum_{i=1}^{n} \beta_i B_i(x, y) \right]^2 dxdy. \tag{6.17}$$

Using essentially the same analysis as we did in Section 6.1, we can prove the following characterization result.

THEOREM 6.4

The least squares variational problem of finding $\boldsymbol{\beta}^*$ satisfying (6.17) has a unique solution,

$$\sum_{i=1}^{n} \beta_i^* B_i(x, y),$$

where the coefficients $\boldsymbol{\beta}^*$ are the solution of the symmetric, positive definite linear system

$$A\boldsymbol{\beta}^* = \mathbf{k}, \tag{6.18}$$

where

$$A \equiv [a_{ij}] \equiv \left[\int_0^1 \int_0^1 B_i(x, y) B_j(x, y) \, dxdy \right]$$

and

$$\mathbf{k} \equiv [k_i] \equiv \left[\int_0^1 \int_0^1 f(x, y) B_i(x, y) \, dxdy \right].$$

If $S \equiv \left\{ \sum_{i=1}^{n} \beta_i B_i(x, y) \mid \boldsymbol{\beta} \in R^n \right\}$, then the unique solution given by (6.18) is denoted by $P_S f \equiv \sum_{i=1}^{n} \beta_i^* B_i(x, y)$. Clearly $P_S f$ is the orthogonal projection of f onto S with respect to the L^2-inner product over the square U; cf. [6.4].

We now examine the choice of basis functions. As a basis for $L(\rho)$, we suggest the functions $\{l_i(x) l_j(y)\}_{i=0, j=0}^{N+1, M+1}$. Using these to form the matrix of (6.18), we obtain a sparse matrix with nine nonzero diagonals. The zero structure is given in the following figure

$$A_{L(\rho)} = \begin{bmatrix} \ddots & & \ddots & 0 \\ & \ddots & 0 & \ddots \\ \ddots & 0 & \ddots & \\ 0 & \ddots & & \ddots \end{bmatrix},$$

where the solid lines indicate the nonzero diagonals.

Actually $A_{L(\rho)}$ may be conveniently expressed in terms of $A_{L(\Delta)}$ and $A_{L(\Delta_y)}$. In fact, if $B \equiv [b_{ij}]$ and $C \equiv [c_{ij}]$ are an $n \times n$ and $m \times m$ matrix respectively, then we define their tensor or Kronecker product as the $mn \times mn$ matrix $B \otimes C$ given by

$$B \otimes C \equiv \begin{bmatrix} b_{11}C & \cdots & b_{1n}C \\ \cdot & & \cdot \\ \cdot & & \cdot \\ \cdot & & \cdot \\ b_{n1}C & \cdots & b_{nn}C \end{bmatrix},$$

and in particular $A_{L(\rho)} = A_{L(\Delta)} \otimes A_{L(\Delta_y)}$. Moreover, it follows from the theory of tensor products of matrices (cf. [6.1]), that

$$\text{cond}\,(A_{L(\rho)}) \leq 36\underline{h}^{-1}h\underline{k}^{-1}k.$$

To solve the linear system corresponding to $A_{L(\rho)}$, we suggest either Cholesky decomposition (cf. [6.2] and [6.6]), or an iterative method such as successive overrelaxation (SOR); cf. [6.6] and [6.11]. The Cholesky decomposition requires storage of the order $\underline{\rho}^{-3}$ and on the order of $\underline{\rho}^{-3}$ arithmetic operations. The successive overrelaxation iterative method requires storage of the order $\underline{\rho}^{-1}$ and by Ostrowski's Theorem is convergent for any relaxation factor $\omega \in (0, 2)$; cf. [6.11]. Moreover, if we consider the block partitioned form of $A_{L(\rho)}$ obtained by lumping together all the unknowns along every horizontal line (i.e., for each $0 \leq j \leq M + 1$, we consider as one vector in R^{N+2} the coefficients of $l_i(x)l_j(y)$, $0 \leq i \leq N + 1$), we obtain a block tridiagonal matrix $\tilde{A}_{L(\rho)}$. Since $\tilde{A}_{L(\rho)}$ has block property A and is block consistently ordered, we may apply the theory of Young (cf. [6.11]) to determine the optimal relaxation factor for block successive relaxation.

As a basis for $H(\rho)$, we suggest the functions

$$\{h_i(x)h_j(y), \tilde{h}_i^1(x)h_j(y), h_i(x)\tilde{h}_j^1(y), \tilde{h}_i^1(x)\tilde{h}_j^1(y) \,|\, 0 \leq i \leq N + 1 \text{ and } 0 \leq j \leq M + 1\}. \tag{6.19}$$

If for each i and j, we lump together the four basis functions $h_i(x)h_j(y)$, $\tilde{h}_i^1(x)h_j(y)$, $h_i(x)\tilde{h}_j^1(y)$, and $\tilde{h}_i^1(x)\tilde{h}_j^1(y)$, then we obtain a block matrix whose zero structure is the same as the zero structure of the preceding matrix $A_{L(\rho)}$. Moreover, for uniform partitions, ρ, cond $A_{H(\rho)} \leq (160)^2$. As discussed before, we may use either Cholesky decomposition or successive overrelaxation to solve the corresponding system.

Finally, as a basis for $S(\rho)$, we suggest the functions

$$\{s_i(x)s_j(y) \,|\, -3 \leq i \leq N \text{ and } -3 \leq j \leq M\}, \tag{6.20}$$

where $s_i(x)$ and $s_j(y)$ are defined in (6.16). Using these to form the least squares matrix (6.18), we obtain a sparse matrix with forty-nine nonzero diagonals. Again the corresponding linear system can be solved by either Cholesky decomposition or successive overrelaxation.

6.3 ERROR ANALYSIS

In this section, we prove a priori error bounds for the least squares procedures introduced in Sections 6.1 and 6.2. Our analysis is based upon the results of Chapters 2, 3, and 4.

THEOREM 6.5

If $f \in PC^{2,2}(I)$, then

$$\|f - P_{L(\Delta)} f\|_2 \leq \pi^{-2} h^2 \|D^2 f\|_2 \tag{6.21}$$

and

$$\|D(f - P_{L(\Delta)} f)\|_2 \leq \pi^{-1}(1 + 4\sqrt{3}\pi^{-1} h^{-1} h) h \|D^2 f\|_2. \tag{6.22}$$

Proof. Inequality (6.21) follows from the observation that

$$\|f - P_{L(\Delta)} f\|_2 = \inf_{l(x) \in L(\Delta)} \|f - l\|_2 \leq \|f - \vartheta_{L(\Delta)} f\|_2 \tag{6.23}$$

and the results of Theorem 2.5, which we use to bound the right-hand side of (6.23). To prove (6.22), we use the Schmidt Inequality (cf. Theorem 1.5), and (6.23) to obtain

$$\begin{aligned}
\|D(f - P_{L(\Delta)} f)\|_2 &\leq \|D(f - \vartheta_{L(\Delta)} f)\|_2 + \|D(\vartheta_{L(\Delta)} f - P_{L(\Delta)} f)\|_2 \\
&\leq \|D(f - \vartheta_{L(\Delta)} f)\|_2 \\
&\quad + 2\sqrt{3}\underline{h}^{-1} \|\vartheta_{L(\Delta)} f - P_{L(\Delta)} f\|_2 \\
&\leq \|D(f - \vartheta_{L(\Delta)} f)\|_2 \\
&\quad + 2\sqrt{3}\underline{h}^{-1}(\|f - \vartheta_{L(\Delta)} f\|_2 + \|f - P_{L(\Delta)} f\|_2) \\
&\leq \|D(f - \vartheta_{L(\Delta)} f)\|_2 + 4\sqrt{3}\underline{h}^{-1} \|f - \vartheta_{L(\Delta)} f\|_2.
\end{aligned} \tag{6.24}$$

Inequality (6.22) now follows by using the results of Theorem 2.5 to bound the right-hand side of (6.24) Q.E.D.

In the L^∞-norm, we can obtain a rather surprising error bound, which states that the least squares approximation to $f \in PC^{2,\infty}(I)$ is asymptotically as good as the Tchebyscheff approximation to f in $L(\Delta)$.

THEOREM 6.6

If $f \in PC^{2,\infty}(I)$, then

$$\|f - P_{L(\Delta)}f\|_\infty \leq \tfrac{1}{12}(1 + 4\underline{h}^{-1}h)h^2 \|D^2 f\|_\infty \tag{6.25}$$

and

$$\|D(f - P_{L(\Delta)}f)\|_\infty \leq \tfrac{1}{2}(1 + \underline{h}^{-2}h^2)h \|D^2 f\|_\infty. \tag{6.26}$$

Proof. The idea of the proof is to show that $P_{L(\Delta)}$ is the second derivative of the cubic spline interpolate of a second iterated integral of $f(x)$. The required inequalities then follow immediately from Exercises (4.2) and (4.3).

We define a mapping M of $C[0, 1]$ into $C^2[0, 1]$ by

$$M(f)(x) \equiv \int_0^x \int_0^s f(t)\, dt ds.$$

It is easy to verify directly that $M(f)(0) = DM(f)(0) = 0$, $M\{L(\Delta)\} = \{s \in S(\Delta) \mid s(0) = Ds(0) = 0\} \equiv \tilde{S}(\Delta)$, and $D^2 M(f) = f$. Moreover, since from (6.3) we have $(f - P_{L(\Delta)}f, l)_2 = 0$ for all $l \in L(\Delta)$,

$$(D^2 M(f) - D^2 M(P_{L(\Delta)}f), D^2 M(l))_2 = 0 \qquad \text{for all } l \in L(\Delta), \tag{6.27}$$

or equivalently

$$(D^2 M(f) - D^2 M(P_{L(\Delta)}f), D^2 s)_2 = 0 \qquad \text{for all } s \in \tilde{S}(\Delta). \tag{6.28}$$

But from the proof of Theorem 4.3, we have

$$(D^2 M(f) - D^2 \vartheta_{S(\Delta)} M(f), D^2 s)_2 = 0 \qquad \text{for all } s \in S(\Delta), \tag{6.29}$$

and subtracting (6.28) from (6.29) we obtain

$$(D^2 M(P_{L(\Delta)}f) - D^2 \vartheta_{S(\Delta)} M(f), D^2 s)_2 = 0 \qquad \text{for all } s \in \tilde{S}(\Delta). \tag{6.30}$$

Since $[M(P_{L(\Delta)}f) - \vartheta_{S(\Delta)} M(f)] \in \tilde{S}(\Delta)$, we have by the "one-sided Rayleigh–Ritz Inequality" (cf. Exercise (1.3)),

$$\begin{aligned} 0 &= \|D^2[M(P_{L(\Delta)}f) - \vartheta_{S(\Delta)} M(f)]\|_2^2 \\ &\geq \left(\frac{\pi^2}{4}\right)^2 \|M(P_{L(\Delta)}f) - \vartheta_{S(\Delta)} M(f)\|_2^2, \end{aligned} \tag{6.31}$$

or $M(P_{L(\Delta)}f) = \vartheta_{S(\Delta)} M(f)$. Hence,

$$P_{L(\Delta)}f = D^2 M(P_{L(\Delta)}f) = D^2 \vartheta_{S(\Delta)} M(f),$$

and

$$f - P_{L(\Delta)} f = D^2 M(f) - D^2 \vartheta_S M(f).$$

The results now follow by applying Exercises (4.2) and (4.3) to the function $M(f)$. Q.E.D.

It is possible to use the argument of the preceding proof in reverse to obtain new error bounds in the L^2-norm for cubic spline interpolation from the results of Theorem 6.5. We do this now to improve and extend some of the results of Chapter 4. We begin by improving the results of Theorems 4.6 and 4.9.

THEOREM 6.7

If $f \in PC^{4,2}(I)$, then

$$\| D^2(f - \vartheta_S f) \|_2 \leq \pi^{-2} h^2 \| D^4 f \|_2, \tag{6.32}$$

$$\| D(f - \vartheta_S f) \|_2 \leq 2\pi^{-3} h^3 \| D^4 f \|_2, \tag{6.33}$$

and

$$\| f - \vartheta_S f \|_2 \leq 2\pi^{-4} h^4 \| D^4 f \|_2. \tag{6.34}$$

Proof. We first remark that

$$f(x) - [Df(0)x + f(0)] - \vartheta_S[f(x) - (Df(0)x + f(0))] = f(x) - \vartheta_S f(x),$$

since ϑ_S preserves cubic polynomials. Thus, we may assume that $f(0) = Df(0) = 0$. By the argument of the preceding proof, $D^2 \vartheta_S f = P_{L(\Delta)} D^2 f$ and hence by inequality (6.21)

$$\| D^2(f - \vartheta_S f) \|_2 = \| D^2 f - P_{L(\Delta)} D^2 f \|_2 \leq \pi^{-2} h^2 \| D^4 f \|_2,$$

which proves (6.32). The remaining results follow from (6.32) as in the proof of Theorem 4.6. Q.E.D.

The following result is proved in the same way as Theorem 6.7. It is an improvement of Theorem 4.9.

THEOREM 6.8

If $f \in PC^{4,2}(U)$, then

$$\begin{aligned} \| f - \vartheta_{S(\rho)} f \|_2 &\leq 2\pi^{-4}(h^4 \| D_x^4 f \|_2 + 2h^2 k^2 \| D_x^2 D_y^2 f \| + k^4 \| D_y^4 f \|_2) \\ &\leq 2\pi^{-4} \bar{\rho}^4 (\| D_x^4 f \|_2 + 2 \| D_x^2 D_y^2 f \|_2 + \| D_y^4 f \|_2). \end{aligned} \tag{6.35}$$

In order to extend the results of the preceding theorem to include bounds for the partial derivatives of the interpolation error, we need the following result, which is also of interest for its own sake.

THEOREM 6.9

If $f \in PC^{3,2}(I)$, then

$$\| D^2(f - \vartheta_S f) \|_2 \leq \pi^{-1} h \| D^3 f \|_2, \tag{6.36}$$

$$\| D(f - \vartheta_S f) \|_2 \leq 2\pi^{-2} h^2 \| D^3 f \|_2, \tag{6.37}$$

and

$$\| f - \vartheta_S f \|_2 \leq 2\pi^{-3} h^3 \| D^3 f \|_2. \tag{6.38}$$

Proof. As in the proof of Theorem 6.7, we may assume that $f(0) = Df(0) = 0$ and hence $D^2 \vartheta_S f = P_{L(\Delta)} D^2 f$. Thus, we have

$$\| D^2(f - \vartheta_S f) \|_2 = \| D^2 f - P_{L(\Delta)} D^2 f \|_2,$$

and inequality (6.36) follows from Exercise (6.1). The remaining inequalities follow from (6.36) as in the proof of Theorem 4.5. Q.E.D.

THEOREM 6.10

If $f \in PC^{4,2}(U)$, then

$$\begin{aligned} &\| D_x(f - \vartheta_{S(\rho)} f) \|_2 \\ &\quad \leq 2\pi^{-3}(h^3 \| D_x^4 f \|_2 + 2hk^2 \| D_x^2 D_y^2 f \|_2 + k^3 \| D_x D_y^3 f \|_2) \\ &\quad \leq 2\pi^{-3} \bar{\rho}^3 (\| D_x^4 f \|_2 + 2 \| D_x^2 D_y^2 f \|_2 + \| D_x D_y^3 f \|_2) \end{aligned} \tag{6.39}$$

and

$$\begin{aligned} &\| D_y(f - \vartheta_{S(\rho)} f) \|_2 \\ &\quad \leq 2\pi^{-3}(k^3 \| D_y^4 f \|_2 + 2kh^2 \| D_x^2 D_y^2 f \|_2 + h^3 \| D_y D_x^3 f \|_2) \\ &\quad \leq 2\pi^{-3} \bar{\rho}^3 (\| D_y^4 f \|_2 + 2 \| D_x^2 D_y^2 f \|_2 + \| D_y D_x^3 f \|_2). \end{aligned} \tag{6.40}$$

Proof. We prove only (6.39), since (6.40) follows by symmetry. Using the results of Theorems 4.5, 6.7, and 6.9, we have

$$\begin{aligned} \| D_x(f - \vartheta_{S(\rho)} f) \|_2 &\leq \| D_x(f - \vartheta_{S(\Delta)} f) \|_2 + \| D_x[\vartheta_{S(\Delta)}(f - \vartheta_{S(\Delta_y)} f) \\ &\qquad - (f - \vartheta_{S(\Delta_y)} f)] \|_2 + \| D_x(f - \vartheta_{S(\Delta_y)} f) \|_2 \\ &\leq 2\pi^{-3} h^3 \| D_x^4 f \|_2 + 2\pi^{-1} h \| D_x^2(f - \vartheta_{S(\Delta_y)} f) \|_2 \\ &\qquad + 2\pi^{-3} k^3 \| D_x D_y^3 f \|_2 \\ &\leq 2\pi^{-3} h^3 \| D_x^4 f \|_2 + 4\pi^{-3} hk^2 \| D_x^2 D_y^2 f \|_2 \\ &\qquad + 2\pi^{-3} k^3 \| D_x D_y^3 f \|_2. \end{aligned} \tag{6.41}$$

Q.E.D.

We now return to the question of error bounds for finite element regression. Using the same techniques as in the proof of Theorem 6.5, we may obtain the following result.

THEOREM 6.11

If $f \in PC^{4,2}(I)$, then

$$\|f - P_{H(\Delta)}f\|_2 \leq \pi^{-4}h^4\|D^4f\|_2, \tag{6.42}$$

$$\|D(f - P_{H(\Delta)}f)\|_2 \leq \pi^{-3}(1 + 2\sqrt{90 + 2\sqrt{1605}}\pi^{-1}\underline{h}^{-1}h)h^3\|D^4f\|_2 \tag{6.43}$$

and

$$\|D^2(f - P_{H(\Delta)}f)\|_2 \leq \pi^{-2}(1 + 4\sqrt{1350 + 30\sqrt{1605}}\pi^{-2}\underline{h}^{-2}h^2)\|D^4f\|_2. \tag{6.44}$$

Using a result of Hall (cf. [6.5]) concerning $C^3(I)$-piecewise quintic polynomial interpolation of functions in $PC^{6,\infty}$, we may prove the following analogue of the surprising result of Theorem 6.6.

THEOREM 6.12

If $f \in PC^{4,\infty}(I)$, then

$$\|f - P_{H(\Delta)}f\|_\infty \leq \tfrac{11}{5760}h^4\|D^4f\|_\infty, \tag{6.45}$$

$$\|D(f - P_{H(\Delta)}f)\|_\infty \leq \tfrac{1}{120}(1 + 2\underline{h}^{-1}h)h^3\|D^4f\|_\infty, \tag{6.46}$$

and

$$\|D^2(f - P_{H(\Delta)}f)\|_\infty \leq \tfrac{1}{60}(6 + 5\underline{h}^{-2}h^2)h^2\|D^4f\|_\infty. \tag{6.47}$$

Turning now to the spline case, we can prove the following result with the aid of Theorem 6.7.

THEOREM 6.13

If $f \in PC^{4,2}(I)$, then

$$\|f - P_{S(\Delta)}f\|_2 \leq 2\pi^{-4}h^4\|D^4f\|_2, \tag{6.48}$$

$$\|D(f - P_{S(\Delta)}f)\|_2 \leq 2\pi^{-3}(1 + 2\sqrt{90 + 2\sqrt{1605}}\pi^{-1}\underline{h}^{-1}h)h^3\|D^4f\|_2, \tag{6.49}$$

and

$$\|D^2(f - P_{S(\Delta)}f)\|_2 \leq 2\pi^{-2}(1 + 4\sqrt{1350 + 30\sqrt{1605}}\pi^{-2}\underline{h}^{-2}h^2)h^2\|D^4f\|_2. \tag{6.50}$$

To prove an analogue of Theorems 6.6 and 6.12 for $S(\Delta)$, we need a result about L^∞-norm bounds for the error in interpolation in $S(5, \Delta, 4)$, where we have used the notations of Exercise (4.12). More precisely, we need to know that there exist positive constants, $\{\mu_k\}_{k=0}^4$, such that

$$\|D^k(f - \vartheta_3 f)\|_\infty \leq \mu_k h^{6-k} \|D^6 f\|_\infty, \qquad 0 \leq k \leq 4, \tag{6.51}$$

for all $f \in PC^{6,\infty}(I)$; cf. [6.10] for a discussion of such results.

THEOREM 6.14

If $f \in PC^{4,\infty}(I)$ and (6.54) is true, then

$$\|D^j(f - P_{S(\Delta)} f)\|_\infty \leq \mu_{j+2} h^{4-j} \|D^4 f\|_\infty, \qquad 0 \leq j \leq 2. \tag{6.52}$$

We turn now to the derivation of error bounds for two-dimensional problems.

THEOREM 6.15

If $f \in PC^{2,2}(U)$, then

$$\begin{aligned} \|f - P_{L(\rho)} f\|_2 &\leq \pi^{-2}(h^2 \|D_x^2 f\|_2 + k^2 \|D_y^2 f\|_2) \\ &\leq \pi^{-2} \bar{\rho}^2 (\|D_x^2 f\|_2 + \|D_y^2 f\|_2), \end{aligned} \tag{6.53}$$

$$\begin{aligned} &\|D_x(f - P_{L(\rho)} f)\|_2 \\ &\leq \pi^{-1}[(1 + 4\sqrt{3}\,\underline{h}^{-1} h) h \|D_x^2 f\|_2 + (2\sqrt{3}\,\pi^{-1} h^{-1} k) k \|D_y^2 f\|_2] \\ &\leq \pi^{-1}(1 + 4\sqrt{3}\,\underline{\rho}^{-1} \bar{\rho}) \bar{\rho} (\|D_x^2 f\|_2 + \|D_y^2 f\|_2), \end{aligned} \tag{6.54}$$

and

$$\begin{aligned} &\|D_y(f - P_{L(\rho)} f)\|_2 \\ &\leq \pi^{-1}[(2\sqrt{3}\,\pi^{-1} \underline{k}^{-1} h) h \|D_x^2 f\|_2 + (1 + 4\sqrt{3}\,\underline{k}^{-1} k) k \|D_y^2 f\|_2] \\ &\leq \pi^{-1}(1 + 4\sqrt{3}\,\underline{\rho}^{-1} \bar{\rho}) \bar{\rho} (\|D_x^2 f\|_2 + \|D_y^2 f\|_2). \end{aligned} \tag{6.55}$$

Proof. We begin by remarking that

$$P_{L(\rho)} f = P_{L(\Delta)} P_{L(\Delta_y)} f = P_{L(\Delta_y)} P_{L(\Delta)} f.$$

In fact, if $\{A_i(x)\}_{i=0}^{N+1}$ denotes an orthonormal basis for $L(\Delta)$ and $\{F_j(y)\}_{j=0}^{M+1}$ denotes an orthonormal basis for $L(\Delta_y)$, then $\{A_i(x)F_j(y)\}_{i=0,\,j=0}^{N+1,\,M+1}$ is an orthonormal basis for $L(\rho)$ and

$$\begin{aligned} P_{L(\rho)} f &= \sum_{i=0}^{N+1} \sum_{j=0}^{M+1} A_i(x) F_j(y) \int_0^1 \int_0^1 f(x, y) A_i(x) F_j(y) \, dx dy \\ &= \sum_{i=0}^{N+1} A_i(x) \int_0^1 \left[\sum_{j=0}^{M+1} F_j(y) \int_0^1 f(x, y) F_j(y) \, dy \right] A_i(x) \, dx \\ &= P_{L(\Delta)} P_{L(\Delta_y)} f \\ &= \sum_{j=0}^{M+1} F_j(y) \int_0^1 \left[\sum_{i=0}^{N+1} A_i(x) \int_0^1 f(x, y) A_i(x) \, dx \right] F_j(y) \, dy \\ &= P_{L(\Delta_y)} P_{L(\Delta)} f. \end{aligned}$$

Thus, since $\int_0^1 [P_{L(\Delta)}g(x)]^2\,dx \le \int_0^1 [g(x)]^2\,dx$ for all $g(x) \in PC^{0,2}(I)$ (cf. [6.4, p. 175]), we have

$$\begin{aligned} \|f - P_{L(\rho)}f\|_2 &\le \|f - P_{L(\Delta)}f\|_2 + \|P_{L(\Delta)}(f - P_{L(\Delta_y)}f)\|_2 \\ &\le \|f - P_{L(\Delta)}f\|_2 + \|f - P_{L(\Delta_y)}f\|_2. \end{aligned} \tag{6.56}$$

Inequality (6.53) follows by using the results of Theorem 6.5 to bound the right-hand side of (6.56).

To obtain (6.54) we use the Schmidt Inequality (cf. Theorem 1.5), to get

$$\begin{aligned} \|D_x(f - P_{L(\rho)}f)\|_2 &\le \|D_x(f - P_{L(\Delta)}f)\|_2 + \|D_x P_{L(\Delta)}(f - P_{L(\Delta_y)}f)\|_2 \\ &\le \|D_x(f - P_{L(\Delta)}f)\|_2 + 2\sqrt{3}\,\underline{h}^{-1}\|f - P_{L(\Delta_y)}f\|_2. \end{aligned} \tag{6.57}$$

Inequality (6.54) follows by using the results of Theorem 6.5 to bound the right-hand side of (6.57). Inequality (6.55) follows by symmetry. Q.E.D.

In a similar fashion, we may prove the following result.

THEOREM 6.15

If $f \in PC^{4,2}(U)$, then

$$\begin{aligned} \|f - P_{H(\rho)}f\|_2 &\le \pi^{-4}(h^4\|D_x^4 f\|_2 + k^4\|D_y^4 f\|_2) \\ &\le \pi^{-4}\bar{\rho}^4(\|D_x^4 f\|_2 + \|D_y^4 f\|_2) \end{aligned} \tag{6.58}$$

and

$$\begin{aligned} \|f - P_{S(\rho)}f\|_2 &\le 2\pi^{-4}(h^4\|D_x^4 f\|_2 + k^4\|D_y^4 f\|_2) \\ &\le 2\pi^{-4}\bar{\rho}^4(\|D_x^4 f\|_2 + \|D_y^4 f\|_2). \end{aligned} \tag{6.59}$$

Analogues of inequalities (6.54) and (6.55) hold for the spaces $H(\rho)$ and $S(\rho)$. Their derivations are straightforward and are left to the reader.

EXERCISES FOR CHAPTER 6

(6.1) Show that if $f \in PC^{1,2}(I)$, then

$$\|f - P_{L(\Delta)}f\|_2 \le \pi^{-1}h\|Df\|_2.$$

(6.2) Show that $\boldsymbol{\beta}$ solves the variational problem (6.1) if and only if it solves the variational problem

$$\inf_{\boldsymbol{\beta} \in R^n}\left\{\int_0^1\left[\sum_{i=1}^n \beta_i B_i(x)\right]^2 dx - 2\int_0^1 f(x)\sum_{i=1}^n \beta_i B_i(x)\,dx\right\}.$$

(6.3) Using the notations and results of Exercises (4.12)–(4.15), show that if $f \in PC^{2m,2}(I)$, then

$$\|f - P_{S(2m-1,\Delta,z)}f\|_2 \leq K_{m,2m,z,0}\,h^{2m}\,\|D^{2m}f\|_2.$$

Moreover, show that if $f \in PC^{2m,2}(R)$, then

$$\|f - P_{S(2m-1,\Delta,z)\otimes S(2m-1,\Delta_y,z)}f\|_2 \leq K_{m,2m,z,0}(h^{2m}\,\|D_x^{2m}f\|_2 + k^{2m}\,\|D_y^{2m}f\|_2).$$

(6.4) In many problems we wish to approximate a closed contour $\Gamma \equiv \{(x(t), y(t))\,|\,0 \leq t \leq 1\}$ in the plane. We may assume that $(x(0), y(0)) = (0, 0)$ and since Γ is closed we must also have $(x(1), y(1)) = (0, 0)$. If

$$L_0(\Delta) \equiv \{l(x) \in L(\Delta)\,|\,l(0) = l(1) = 0\},$$
$$H_0(\Delta) \equiv \{h(x) \in H(\Delta)\,|\,h(0) = h(1) = 0\},$$

and

$$S_0(\Delta) \equiv \{s(x) \in S(\Delta)\,|\,s(0) = s(1) = 0\},$$

develop analogues of the results of Section 6.1 for the least squares approximation of the two coordinate functions $x(t)$ and $y(t)$ over $L_0(\Delta)$, $H_0(\Delta)$, and $S_0(\Delta)$. Many times we are given $(x(t), y(t))$ for only a discrete set of points $T \equiv \{t_i\}_{i=1}^{Q}$, i.e., we are given only a finite set of points, in the plane, which approximate the contour. In this case develop analogues of the "approximate regression" results of Section 6.3 (Hint: Use the arc length along the polygon with vertices $\{(x(t), y(t))\,|\,t \in T\}$ as the parameter t.).

(6.5) If $f \in PC^{2,2}(R)$, show that

$$\max(\|D_x(f - P_{H(\rho)}f)\|_2, \|D_y(f - P_{H(\rho)}f)\|_2) \leq \pi^{-1}(1 + 2\sqrt{90 + 2\sqrt{1605}}\,\underline{\rho}^{-1}\bar{\rho})\bar{\rho}(\|D_x^2 f\|_2 + \|D_y^2 f\|_2)$$

and

$$\max(\|D_x(f - P_{S(\rho)}f)\|_2, \|D_y(f - P_{S(\rho)}f)\|_2) \leq 2\pi^{-1}(1 + 2\sqrt{90 + 2\sqrt{1605}}\,\underline{\rho}^{-1}\bar{\rho})\bar{\rho}(\|D_x^2 f\|_2 + \|D_y^2 f\|_2).$$

(6.6) If $f \in PC^{4,2}(U)$, show that

$$\max(\|D_x(f - P_{H(\rho)}f)\|_2, \|D_y(f - P_{H(\rho)}f)\|_2) \leq \pi^{-3}(1 + 2\sqrt{90 + 2\sqrt{1605}}\,\underline{\rho}^{-1}\bar{\rho})\bar{\rho}^3(\|D_x^4 f\|_2 + \|D_y^4 f\|_2)$$

and

$$\max(\|D_x(f - P_{S(\rho)}f)\|_2, \|D_y(f - P_{S(\rho)}f)\|_2) \leq 2\pi^{-3}(1 + 2\sqrt{90 + 2\sqrt{1605}}\,\underline{\rho}^{-1}\bar{\rho})\bar{\rho}^3(\|D_x^4 f\|_2 + \|D_y^4 f\|_2).$$

REFERENCES FOR CHAPTER 6

[6.1] BELLMAN, R., *Introduction to Matrix Theory.* McGraw-Hill Book Co., Inc., New York (1960).

[6.2] FORSYTHE, G., and C. B. MOLER, *Computer Solution of Linear Algebraic Equations.* Prentice-Hall, Inc., Englewood Cliffs, N. J. (1967).

[6.3] GILLON, A., and M. H. SCHULTZ, On the conditioning of matrices arising in the finite element method. Yale Computer Science Research Report.

[6.4] GOFFMAN, C., and G. PEDRICK, *First Course in Functional Analysis.* Prentice-Hall, Inc., Englewood Cliffs, N. J. (1965).

[6.5] HALL, C. A., On error bounds for spline interpolation. *J. Approx. Theory* **1**, 209–218 (1968).

[6.6] ISAACSON, E., and H. B. KELLER, *Analysis of Numerical Methods.* John Wiley & Sons, Inc., New York (1966).

[6.7] PATENT, P., *Least Square Polynomial Spline Approximation.* Ph.D. dissertation, California Institute of Technology (June, 1972).

[6.8] SCHOENBERG, I. J., Contributions to the problem of approximation of equidistant data by analytic functions. Parts A and B. *Quart. Appl. Math.* **4**, 45–99, 112–141 (1946).

[6.9] SCHULTZ, M. H., *Polynomial spline regression.* Yale Computer Science Research Report.

[6.10] SWARTZ, K. B., and R. S. VARGA, Bounds for some spline interpolation errors. (to appear).

[6.11] VARGA, R. S., *Matrix Iterative Analysis.* Prentice-Hall, Inc., Englewood Cliffs, N. J. (1962).

7 THE RAYLEIGH–RITZ–GALERKIN PROCEDURE FOR ELLIPTIC DIFFERENTIAL EQUATIONS

7.1 INTRODUCTION

In the past few years there has been renewed interest in the Rayleigh–Ritz–Galerkin procedure, and in particular the finite element procedure, for approximating the solutions of well-posed boundary value problems for linear and nonlinear elliptic differential equations. In this context, the finite element procedure is nothing more than the Rayleigh–Ritz–Galerkin procedure applied to the spaces of piecewise polynomial functions which we introduced in Chapters 2, 3, and 4. For classical accounts of this procedure, see [7.14], [7.15], [7.16], [7.25], and [7.26]; for modern accounts see [7.1], [7.7], [7.8], [7.12], [7.31], [7.32], and [7.35].

For a bibliography of the extensive Russian work on this procedure, see [7.27], and for a bibliography of the engineering literature on the finite element method, see [7.44]. Finally, we mention the very general and important work of Aubin, Babuška, Bramble, Fix, Schatz, and Strang on the mathematics of the finite element method; cf. [7.2], [7.3], [7.6], and [7.42] for references.

In this chapter, we will consider the Dirichlet problem for self-adjoint, second-order linear and semilinear elliptic equations. See [7.9], [7.10], [7.11], [7.12], [7.18], [7.37], [7.38], and [7.39] for a discussion of other types of boundary conditions and more general equations. Moreover, for two-dimensional problems we will consider only tensor product types of subspaces. See [7.19], [7.44], [7.45], [7.46], and [7.47] for discussions of subspaces based on triangulations.

7.2 LINEAR SECOND-ORDER TWO-POINT BOUNDARY VALUE PROBLEMS

In this section, we consider the problem of approximating the solution of the self-adjoint differential equation

$$-D[p(x)Du(x)] + q(x)u(x) = f(x), \qquad 0 < x < 1, \tag{7.1}$$

subject to the Dirichlet boundary conditions

$$u(0) = u(1) = 0. \tag{7.2}$$

We assume that the differential equation is elliptic, i.e., $p(x)$ and $q(x) \in PC^{0,\infty}(I)$ are such that there exist positive constants γ and μ such that

$$\gamma \| Du \|_2^2 \leq \int_0^1 [p(x)(Du)^2 + q(x)u^2]\, dx \leq \mu \| Du \|_2^2 \tag{7.3}$$

for all $u \in PC_0^{1,2}(I) \equiv \{\phi \in PC^{1,2}(I) \mid \phi(0) = \phi(1) = 0\}$, and $f \in PC^{0,2}(I)$. See [7.13] for a discussion of singular two-point boundary value problems. We say that u is a *generalized solution* (over $PC_0^{1,2}$) of (7.1)–(7.2) if and only if $u \in PC_0^{1,2}(I)$ and

$$a(u, v) \equiv \int_0^1 [p(x)DuDv + q(x)uv]\, dx = (f, v)_2 \tag{7.4}$$

for all $v \in PC_0^{1,2}(I)$. Integrating by parts, we can prove the following standard result.

THEOREM 7.1

If u is a classical solution of (7.1)–(7.2), i.e., $u \in C^2(I)$ and satisfies (7.1) pointwise, then it is a generalized solution.

Moreover, if the coefficients of the differential equation are sufficiently smooth, we can show that every generalized solution is a classical solution.

We now state and prove a variational characterization of generalized solutions.

THEOREM 7.2

The function $u(x)$ is the generalized solution of (7.1)–(7.2) if and only if $u(x)$ is the unique solution of the variational problem of finding u such that

$$F[u] = \inf_{w \in PC_0^{1,2}} F[w] \equiv a(w, w) - 2(w, f)_2. \tag{7.5}$$

Proof. First, we show that if u solves the variational problem then it is a generalized solution. In fact, if $\eta \in PC_0^{1,2}(I)$ and $\alpha \in R$,

$$F[u + \alpha\eta] = a(u, u) + \alpha^2 a(\eta, \eta) + 2\alpha a(u, \eta) - 2(u, f)_2 - 2\alpha(\eta, f)_2 \geq F[u],$$

and hence $F[u + \alpha\eta]$ is quadratic in α and has a minimum at $\alpha = 0$ only if $(dF/d\alpha)[u] = 0$. Calculating this latter expression, we obtain $a(u, \eta) = (f, \eta)_2$ for all $\eta \in PC_0^{1,2}(I)$, i.e., u is a generalized solution.

Conversely, if u is a generalized solution and $\eta \in PC_0^{1,2}(I)$, then

$$\begin{aligned} F[\eta] - F[u] &= a(\eta, \eta) - a(u, u) + 2(f, u - \eta)_2 \\ &= a(\eta, \eta) + a(u, u) - 2(f, \eta)_2, \end{aligned}$$

where we have used the equality obtained by putting $v \equiv u$ in (7.4). Using the equality obtained by putting $v \equiv \eta$ in (7.4), we finally obtain

$$\begin{aligned} F[\eta] - F[u] &= a(\eta, \eta) - 2a(u, \eta) + a(u, u) = a(\eta - u, \eta - u) \\ &\geq \gamma^2 \| D(u - \eta) \|_2^2 \geq 0, \end{aligned} \tag{7.6}$$

with equality if and only if $u = \eta$, i.e., u is the unique solution of the variational problem (7.5). Q.E.D.

Let S be any finite-dimensional subspace of $PC_0^{1,2}(I)$ and $\{B_i(x)\}_{i=1}^n$ be a basis for S. The Rayleigh–Ritz procedure is to find an approximation to the generalized solution, u, of (7.1)–(7.2) by determining an element $u_S \in S$, which minimizes $F[u]$ over S. The following result shows that this is a well-defined procedure.

THEOREM 7.3

There is a unique element $u_S \in S$ which minimizes $F[u]$ over S.

Proof. Considering

$$F[\boldsymbol{\beta}] \equiv F\left[\sum_{i=1}^n \beta_i B_i\right] = \sum_{i=1}^n \sum_{j=1}^n \beta_i \beta_j a(B_i, B_i) - 2\sum_{i=1}^n \beta_i (f, B_i)_2$$

as a function of $\boldsymbol{\beta} \in R^n$, it is clear that $F[\boldsymbol{\beta}]$ is quadratic in $\boldsymbol{\beta}$ and hence F has a minimum at $\boldsymbol{\beta}^*$ if and only if

$$\frac{\partial F}{\partial \beta_i}[\boldsymbol{\beta}^*] = 0, \qquad \text{for all } 1 \leq i \leq n, \tag{7.7}$$

and the matrix

$$H \equiv \left[\frac{\partial^2 F}{\partial \beta_i \partial \beta_j}[\boldsymbol{\beta}^*]\right]$$

is positive definite.

Calculating the equations of the system (7.7), we obtain

$$\frac{\partial F}{\partial \beta_i}[\boldsymbol{\beta}^*] = 2 \sum_{j=1}^{n} \beta_j a(B_i, B_j) - 2(f, B_i), \qquad 1 \leq i \leq n, \tag{7.8}$$

or in matrix form

$$A\boldsymbol{\beta}^* = \mathbf{k}, \tag{7.9}$$

where

$$A \equiv [a_{ij}] \equiv [a(B_i, B_j)] \tag{7.10}$$

and

$$\mathbf{k} \equiv [k_i] \equiv [(f, B_i)_2]. \tag{7.11}$$

Clearly, A is symmetric and positive definite. In fact, if $\boldsymbol{\beta} \neq \mathbf{0}$, then

$$\boldsymbol{\beta}^T A \boldsymbol{\beta} = a\Big(\sum_{i=1}^{n} \beta_i B_i, \sum_{i=1}^{n} \beta_i B_i\Big) \geq \gamma \,\| D(\sum_{i=1}^{n} \beta_i B_i) \|_2^2$$

$$\geq \gamma\pi^2 \Big\| \sum_{i=1}^{n} \beta_i B_i \Big\|_2^2 > 0.$$

Thus (7.9) has a unique solution $\boldsymbol{\beta}^*$. Moreover, from (7.8) it is clear that $H = 2A$ and hence $\boldsymbol{\beta}^*$ is the unique minimum of F over R^n. Q.E.D.

It follows from the preceding proof that the Rayleigh–Ritz approximation can be characterized in terms of the solution of a linear system of equations, whose matrix is symmetric and positive definite for *any* choice of basis functions for *any* finite-dimensional subspace of $PC_0^{1,2}(I)$.

The Galerkin procedure is to find an approximation to the generalized solution, u, of (7.1)–(7.2) by determining an element $w_S \in S$ such that

$$a(w_S, B_i) = (f, B_i)_2 \qquad \text{for all } 1 \leq i \leq n. \tag{7.12}$$

If we expand w_S in terms of the basis functions

$$w_S(x) = \sum_{i=1}^{n} \gamma_i B_i(x),$$

then we can see that the coefficients, $\boldsymbol{\gamma}$, satisfy the exact same linear algebraic equations as the coefficients, $\boldsymbol{\beta}^*$, for the Rayleigh–Ritz approximation. Thus, for a problem of this form the Rayleigh–Ritz and Galerkin approximations are identical and will henceforth be called the Rayleigh–Ritz–Galerkin or RRG approximation.

We now show how to construct computationally attractive basis functions for the spaces under consideration.

As a basis for $L_0(\Delta) \equiv \{l(x) \in L(\Delta) \mid l(0) = l(1) = 0\}$, we suggest the functions $\{l_i(x)\}_{i=1}^N$ defined in Chapter 2, i.e., we eliminate the two functions $l_0(x)$ and $l_{N+1}(x)$ from our basis for $L(\Delta)$. This choice of basis leads to an RRG system (7.9) with a tridiagonal matrix. As discussed in Chapter 6, such systems are easily solved by Gaussian elimination.

For the special case of a uniform partition with mesh length h, $p(x) \equiv 1$, and $q(x) \equiv 0$, the RRG matrix is given by

$$A_h \equiv h^{-1} \begin{bmatrix} 2 & -1 & & & 0 \\ -1 & \ddots & \ddots & & \\ & \ddots & \ddots & \ddots & \\ & & \ddots & \ddots & -1 \\ 0 & & & -1 & 2 \end{bmatrix}, \tag{7.13}$$

which is irreducibly diagonally dominant; cf. [7.43]. Moreover, since

$$\sin k\pi(x - h) - 2 \sin k\pi x + \sin k\pi(x + h) + 2(1 - \cos k\pi h) \sin k\pi x = 0,$$

the eigenvalues of A_h are $\{2h^{-1}(1 - \cos \pi kh) \mid 1 \leq k \leq N\}$, $h \equiv (N + 1)^{-1}$, and hence

$$\begin{aligned} \text{cond}(A_h) &= (1 - \cos \pi h)^{-1}(1 - \cos \pi Nh) \approx (\pi^2 h^2)^{-1}(\pi^2 N^2 h^2) \\ &= N^2 \approx h^{-2} \quad \text{as} \quad h \longrightarrow 0. \end{aligned}$$

Furthermore, the matrix $h^{-1}A_h$ is identical to the matrix one obtains from the standard three-point central difference approximation to the differential operator $-D^2$; cf. [7.24].

For the case of nonuniform partitions, we can obtain a similar result, though the proof uses the result of Theorem 6.2 and the Schmidt Inequality (cf. Theorem 1.5).

THEOREM 7.4

If $A_{L(\Delta)}$ is the RRG matrix (7.10) obtained by using the basis functions $\{l_i(x)\}_{i=1}^N$, then

$$\text{cond}(A_{L(\Delta)}) \leq 72\gamma^{-1}\mu\pi^{-2}(\underline{h}^{-1}h)\underline{h}^{-2}. \tag{7.14}$$

Proof. From the Rayleigh–Ritz Inequality (Theorem 1.2) and the proof of Theorem 6.2, we have for all $\boldsymbol{\beta} \neq \mathbf{0}$

$$\begin{aligned} \frac{\pi^2}{6}\gamma \underline{h} \sum_{i=1}^N \beta_i^2 \leq \pi^2 \gamma \left\| \sum_{i=1}^N \beta_i l_i \right\|_2^2 \leq \gamma \left\| D\left(\sum_{i=1}^N \beta_i l_i\right) \right\|_2^2 \\ \leq \boldsymbol{\beta}^T A_{L(\Delta)} \boldsymbol{\beta}. \end{aligned} \tag{7.15}$$

From the Schmidt Inequality (cf. Theorem 1.5), and the proof of Theorem 6.2, we have

$$\boldsymbol{\beta}^T A_{L(\Delta)} \boldsymbol{\beta} \leq 12\underline{h}^{-2}\mu \left\| \sum_{i=1}^{N} \beta_i l_i \right\|_2^2 \leq 12\mu h^{-2} h \sum_{i=1}^{N} \beta_i^2. \tag{7.16}$$

Combining (7.15) and (7.16), we obtain (7.14). Q.E.D.

The important thing to note about the bound (7.14) is that the condition number grows at a rate no worse than the square of the size of the system.

As a basis for $H_0(\Delta) \equiv \{h(x) \in H(\Delta) \mid h(0) = h(1) = 0\}$, we suggest the basis functions $\{h_i(x)\}_{i=1}^{N} \cup \{\tilde{h}_i^1(x)\}_{i=0}^{N+1}$ defined in Section 6.1, i.e.,

$$\dim H_0(\Delta) = 2N + 2.$$

This choice yields an RRG system with a block tridiagonal matrix as discussed in Section 6.1 for the corresponding least squares matrix. For a uniform partition with mesh length h, $p(x) \equiv 1$, and $q(x) \equiv 0$, the RRG matrix is given by the following matrix, equation (7.17),

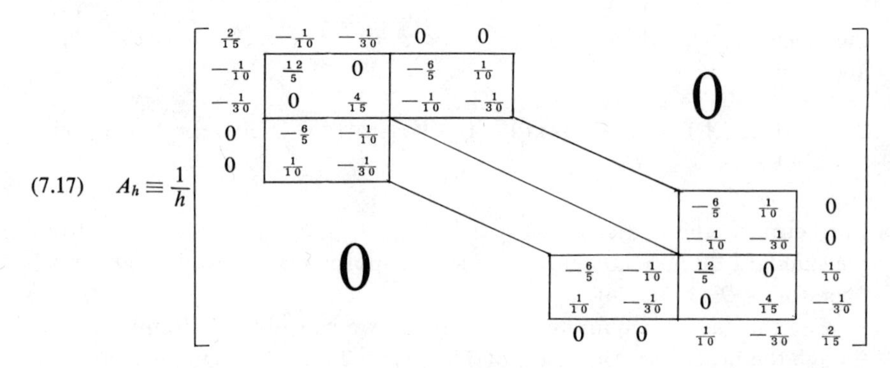

Using the result of Theorem 6.3, we can prove a result on the condition number of $A_{H_0(\Delta)}$ for uniform partitions. See [7.20] for the proof and further details.

THEOREM 7.5

There exists a positive constant, K, such that if Δ is a uniform partition and $A_{H_0(\Delta)}$ is the RRG matrix obtained by using the basis functions $\{h_i(x)\}_{i=1}^{N} \cup \{\tilde{h}_i^1(x)\}_{i=0}^{N+1}$, then

$$\text{cond}\,(A_{H_0(\Delta)}) \leq K\underline{h}^{-2}. \tag{7.18}$$

We make the important observation that the condition number for $H_0(\Delta)$ grows at the same rate as the condition number for $L_0(\Delta)$ even though we have a fourth-order scheme instead of a second-order scheme!

To construct a basis for $S_0(\Delta) \equiv \{s(x) \in S(\Delta) \mid s(0) = s(1) = 0\}$, it is necessary to modify the basis functions $\{s_i(x)\}_{i=-3}^{N}$ presented in Section 6.1 so that the modified functions satisfy the Dirichlet boundary conditions. To this end, we illustrate the procedure by giving the modification for the special case of a uniform mesh.

Let $\tilde{s}_{-2}(x) \equiv s_{-2}(x) - 4s_{-3}(x)$, $\tilde{s}_{-1}(x) \equiv s_{-1}(x) - s_{-3}(x)$, $\tilde{s}_i(x) \equiv s_i(x)$, $0 \leq i \leq N-3$, $\tilde{s}_{N-2}(x) \equiv s_{N-2}(x) - s_N(x)$, and $\tilde{s}_{N-1}(x) \equiv s_{N-1}(x) - 4s_N(x)$. Then $\{\tilde{s}_i(x)\}_{i=-2}^{N-1}$ is a basis for $S_0(\Delta)$ and the support of each $\tilde{s}_i$ is contained in at most four adjacent subintervals of Δ. Thus, use of these basis functions yields a seven-diagonal band RRG matrix, and the corresponding RRG system may be solved by Gaussian elimination.

We turn now to the question of generating the RRG system of linear algebraic equations. The problem is that the nonzero coefficients of the equations are given by integrals of products of the coefficients of the differential equation and the basis functions or their derivatives. In general, these integrals cannot be evaluated analytically, and furthermore we want an automatic program even for those problems in which the integrals can be evaluated analytically. If the coefficients of the differential equation are sufficiently smooth, we suggest approximating the integrals by interpolating the coefficients in the space over which we are doing the RRG procedure and then evaluating the integrals of the approximate integrands, which will be products of piecewise polynomials, exactly. See [7.21], [7.22], [7.40], and [7.41] for further details.

7.3 SEMILINEAR SECOND-ORDER TWO-POINT BOUNDARY VALUE PROBLEMS

In this section, we follow [7.9] and extend the ideas of the previous section to semilinear problems of the form

$$-D[p(x)Du] + q(x)u = f(x, u), \qquad 0 < x < 1, \tag{7.19}$$

subject to the Dirichlet boundary conditions

$$u(0) = u(1) = 0, \tag{7.20}$$

where p and $q \in PC^{0,\infty}$ and are such that there exist positive constants γ and μ such that (7.3) holds for all $u \in PC_0^{1,2}$, $f(x, u)$ and $\frac{\partial f}{\partial u}(x, u)$ are continuous on $[0, 1] \times (-\infty, \infty)$, $\left|\frac{\partial f}{\partial u}(x, u)\right| \leq B$ for all $(x, u) \in [0, 1] \times (-\infty, \infty)$, and

$$\frac{\partial f}{\partial u}(x, u) \leq \lambda < \Lambda \equiv \inf_{w \in PC_0{}^{1,2}(I)} \frac{a(w, w)}{(w, w)_2}.$$

We say that u is a *generalized solution* (over $PC_0^{1,2}$) of (7.19)–(7.20) if and only if $u \in PC_0^{1,2}$ and

$$a(u, v) = (f(u), v)_2, \qquad \text{for all } v \in PC_0^{1,2}. \tag{7.21}$$

Integrating by parts, we can prove the following result.

THEOREM 7.6

If u is a classical solution of (7.19)–(7.20), then it is a generalized solution.

Moreover, if the coefficients of the differential equation are sufficiently smooth, every generalized solution is a classical solution.

As in the linear case, we have uniqueness of generalized solutions.

THEOREM 7.7

The problem (7.19)–(7.20) has at most one generalized solution.

Proof. Let u and v be two distinct generalized solutions. Then

$$\begin{aligned}
0 &= a(u - v, u - v) - (f(u) - f(v), u - v)_2 \\
&= a(u - v, u - v) - \left(\frac{\partial f}{\partial u}(u - v), u - v\right)_2 \\
&\geq a(u - v, u - v) - \lambda(u - v, u - v)_2 \qquad (7.22)\\
&\geq a(u - v, u - v) - \lambda\Lambda^{-1}a(u - v, u - v) \\
&= (1 - \lambda\Lambda^{-1})a(u - v, u - v) \geq \gamma(1 - \lambda\Lambda^{-1})\| D(u - v)\|_2^2 \\
&\geq \gamma\pi^2(1 - \lambda\Lambda^{-1})\| u - v\|_2^2 > 0,
\end{aligned}$$

which is a contradiction. Q.E.D.

We now state and prove a variational characterization of generalized solutions. This is a semilinear generalization of Theorem 7.2.

THEOREM 7.8

The function $u(x)$ is the generalized solution of (7.19)–(7.20) if and only if $u(x)$ solves the variational problem of finding u such that

$$F[u] = \inf_{w \in PC_0^{1,2}} F[w] \equiv \int_0^1 \left[p(x)(Dw)^2 + q(x)w^2 - 2\int_0^{w(x)} f(x, t)\, dt \right] dx. \tag{7.23}$$

Proof. Let $u(x)$ be a generalized solution and $\eta(x)$ be any other element in $PC_0^{1,2}$. Then

$$F[\eta] - F[u] = a(\eta, \eta) - a(u, u) + 2\int_0^1 \left\{\int_0^u f(x, t)\, dt - \int_0^\eta f(x, t)\, dt\right\} dx$$
$$= a(\eta, \eta) - a(u, u) + 2\int_0^1 \int_\eta^u f(x, t)\, dtdx.$$

But since u is a generalized solution, $a(u, u) = (f(u), u)_2$ and $a(u, \eta) = (f(u), \eta)_2$, and hence

$$\begin{aligned} F[\eta] &- F[u] \\ &= a(\eta, \eta) + a(u, u) - 2a(u, \eta) + 2\int_0^1 \int_\eta^u [f(x, t) - f(x, u)]\, dtdx \\ &= a(u - \eta, u - \eta) + 2\int_0^1 \int_\eta^u [f(x, t) - f(x, u)]\, dtdx \\ &= a(u - \eta, u - \eta) - 2\int_0^1 \int_u^\eta \left(\frac{\partial f}{\partial u}\right)(t - u)\, dtdx \qquad (7.24) \\ &\geq a(u - \eta, u - \eta) - \lambda\int_0^1 (\eta - u)^2\, dx \\ &= a(u - \eta, u - \eta) - \lambda \| u - \eta \|_2^2 \geq (1 - \lambda\Lambda^{-1})a(u - \eta, u - \eta) \\ &\geq (1 - \lambda\Lambda^{-1})\gamma\pi^2 \| u - \eta \|_2^2 > 0. \end{aligned}$$

Conversely, if $\eta \in PC_0^{1,2}$ and $\alpha \in R$,

$$F[u + \alpha\eta] = a(u, u) + \alpha^2 a(\eta, \eta) + 2\alpha a(u, \eta) - 2\int_0^1 \int_0^{u+\alpha\eta} f(x, t)\, dtdx$$

is twice continuously differentiable with respect to α and has a minimum at $\alpha = 0$ only if $(dF/d\alpha)[u] = 0$. Calculating this latter expression, we obtain $a(u, \eta) = (f(u), \eta)_2$ for all $\eta \in PC_0^{1,2}$, i.e., u is a generalized solution. Q.E.D.

As in Section 7.1, we let S be any finite-dimensional subspace of $PC_0^{1,2}(I)$ and $\{B_i(x)\}_{i=1}^n$ be a basis for S. The Rayleigh–Ritz procedure is to find an approximation to the generalized solution, u, of (7.19)–(7.20) by determining an element $u_S \in S$ which minimizes $F[w]$ over S. The following result shows that this is a well-defined procedure. Our proof closely follows [7.9].

THEOREM 7.9

There is a unique element $u_S \in S$ which minimizes $F[w]$ over S.

Proof. If $u \in S$, then from Theorem 7.8, u uniquely minimizes F over S. If $u \notin S$, we consider the $(n + 1)$-dimensional space spanned by $\{B_i(x)\}_{i=1}^n$ and u. Any function in this space is expressible as

$$\alpha u(x) + \sum_{i=1}^n \beta_i B_i(x)$$

for suitable coefficients and can be represented as an $(n+1)$-vector $\boldsymbol{\phi} \equiv (\alpha, \beta_1, \ldots, \beta_n)$. Moreover, since $\{B_i(x)\}_{i=1}^n$ and $u(x)$ are linearly independent,

$$|\boldsymbol{\phi}| \equiv |(\alpha, \beta_1, \ldots, \beta_n)| \equiv \left\| \alpha u + \sum_{i=1}^{n} \beta_i B_i \right\|_2 \tag{7.25}$$

is a norm over this subspace, and using (7.24) we have

$$F\left[\sum_{i=1}^{n} \beta_i B_i\right] \geq F[u] + (1 - \lambda\Lambda^{-1})\gamma\pi^2 |(-1, \beta_1, \ldots, \beta_n)|^2. \tag{7.26}$$

But, as all norms on any $(n+1)$-dimensional vector space are equivalent (cf. [7.24]), there exists a positive constant, K, depending on S, such that

$$K|(\alpha, \beta_1, \ldots, \beta_n)| \geq |(\alpha, \beta_1, \ldots, \beta_n)|_2 \geq |(\beta_1, \ldots, \beta_n)|_2,$$

and hence

$$F\left[\sum_{i=1}^{n} \beta_i B_i\right] \geq F[u] + (1 - \lambda\Lambda^{-1})\gamma\pi^2 K^{-2} |(\beta_1, \ldots, \beta_n)|_2^2, \tag{7.27}$$

where $|\cdot|_2$ denotes the l^2-norm. Thus, if we view

$$G(\boldsymbol{\beta}) \equiv G(\beta_1, \ldots, \beta_n) \equiv F\left[\sum_{i=1}^{n} \beta_i B_i\right]$$

as a functional on R^n, then the equivalence of all norms on R^n coupled with (7.27) gives us that

$$\lim_{\|\boldsymbol{\beta}\| \to \infty} G(\boldsymbol{\beta}) = +\infty \tag{7.28}$$

for any norm $\|\cdot\|$ on R^n. Hence, as $G(\boldsymbol{\beta})$ is clearly a continuous function on R^n which is bounded below by $F[u]$ and satisfies (7.28), a standard compactness argument shows that there exists at least one vector $\boldsymbol{\beta}^* \in R^n$ for which $G(\boldsymbol{\beta}) \geq G(\boldsymbol{\beta}^*)$ for all $\boldsymbol{\beta} \in R^n$, or equivalently

$$F\left[\sum_{i=1}^{n} \beta_i B_i\right] \geq F\left[\sum_{i=1}^{n} \beta_i^* B_i\right] \qquad \text{for all } \boldsymbol{\beta} \in R^n. \tag{7.29}$$

To show that $\boldsymbol{\beta}^*$ is unique, we observe that $G(\boldsymbol{\beta})$ is twice continuously differentiable over R^n with derivatives given by

$$\frac{\partial G(\boldsymbol{\beta})}{\partial \beta_i} = 2a\left(\sum_{j=1}^{n} \beta_j B_j, B_i\right) + 2\left(f\left(\sum_{j=1}^{n} \beta_j B_j\right), B_i\right)_2, \qquad 1 \leq i \leq n, \tag{7.30}$$

and

$$\frac{\partial^2 G(\boldsymbol{\beta})}{\partial \beta_i \partial \beta_j} = 2a(B_i, B_j) + 2\left(\frac{\partial f}{\partial u}\left(\sum_{i=1}^{n} \beta_i B_i\right) B_j, B_i\right)_2, \qquad 1 \leq i, j \leq n. \tag{7.31}$$

If we define the $n \times n$ matrix $B(\boldsymbol{\beta}) = [b_{ij}(\boldsymbol{\beta})]$ where $b_{ij}(\boldsymbol{\beta}) \equiv \partial^2 G(\boldsymbol{\beta})/\partial\beta_i\partial\beta_j$ then B is symmetric and uniformly positive definite. In fact,

$$\mathbf{y}^T B(\boldsymbol{\beta})\mathbf{y} = \sum_{i,j=1}^{N} y_i b_{ij}(\boldsymbol{\beta}) y_j = 2a(Y, Y) + 2\left(\frac{\partial f}{\partial u}(\boldsymbol{\beta})Y, Y\right)_2,$$

where $Y(x) \equiv \sum_{i=1}^{n} y_i B_i(x)$. But from the proof of Theorem 7.8, we have

$$\mathbf{y}^T B(\boldsymbol{\beta})\mathbf{y} \geq 2\gamma\pi^2(1 - \lambda\Lambda^{-1})\| Y \|_2^2, \tag{7.32}$$

and, by the equivalence of all norms on R^n,

$$\mathbf{y}^T B(\boldsymbol{\beta})\mathbf{y} \geq 2K^{-2}\gamma\pi^2(1 - \lambda\Lambda^{-1})|\mathbf{y}|_2^2. \tag{7.33}$$

Since $G(\boldsymbol{\beta})$ is twice continuously differentiable, we can write its Taylor series expansion about $\boldsymbol{\beta}^*$ as

$$G(\boldsymbol{\beta}) = G(\boldsymbol{\beta}^*) + (\boldsymbol{\beta} - \boldsymbol{\beta}^*)^T (\text{grad } G(\boldsymbol{\beta}^*)) + (\boldsymbol{\beta} - \boldsymbol{\beta}^*)^T B(\mathbf{w})(\boldsymbol{\beta} - \boldsymbol{\beta}^*), \tag{7.34}$$

where $\mathbf{w} \equiv \theta\boldsymbol{\beta} + (1 - \theta)\boldsymbol{\beta}^*$ for some $\theta \in (0, 1)$. Then uniqueness follows since grad $G(\boldsymbol{\beta}^*) = 0$ implies

$$\begin{aligned} G(\boldsymbol{\beta}) &= G(\boldsymbol{\beta}^*) + (\boldsymbol{\beta} - \boldsymbol{\beta}^*)^T B(\mathbf{w})(\boldsymbol{\beta} - \boldsymbol{\beta}^*) \\ &\geq G(\boldsymbol{\beta}^*) + 2K\gamma\pi^2(1 - \lambda\Lambda^{-1})|\boldsymbol{\beta} - \boldsymbol{\beta}^*|_2^2. \end{aligned}$$

Q.E.D.

To find the unique element $u_S(x) = \sum_{i=1}^{n} \beta_i^* B_i(x)$ which minimizes $F[w]$ over S, we must solve the n nonlinear equations

$$a\left(\sum_{j=1}^{n} \beta_j B_j, B_i\right) = \left(f\left(\sum_{j=1}^{n} \beta_j B_j\right), B_i\right)_2, \qquad 1 \leq i \leq n. \tag{7.35}$$

We can rewrite these equations in vector form as

$$A\boldsymbol{\beta} = \mathbf{g}(\boldsymbol{\beta}) \tag{7.36}$$

where $A \equiv [a_{ij}] = [a(B_i, B_j)]$ and

$$\mathbf{g}(\boldsymbol{\beta}) = [g_i(\boldsymbol{\beta})] = \left[\left(f\left(\sum_{j=1}^{n} \beta_j B_j\right), B_i\right)_2\right].$$

As with the linear problem, we can use the Galerkin procedure to find an approximation to the generalized solution, u, of (7.19)–(7.20) by now determining an element $w_S \in S$ such that

$$a(w_S, B_i) = (f(w_S), B_i)_2, \qquad 1 \leq i \leq n. \tag{7.37}$$

If we expand w_S in terms of the basis functions

$$w_S(x) \equiv \sum_{i=1}^{n} \gamma_i B_i(x),$$

then we can see that the coefficients, $\boldsymbol{\gamma}$, satisfy the exact same nonlinear algebraic equations as the coefficients, $\boldsymbol{\beta}^*$, for the Rayleigh–Ritz approximation. Thus, for a problem of this form the Rayleigh–Ritz and Galerkin approximations are identical, and we will refer to the Rayleigh–Ritz–Galerkin (RRG) approximation.

We now turn briefly to the question of algorithms for solving the nonlinear RRG systems (7.36). Several iterative methods can be rigorously applied; cf. [7.29], [7.33], and [7.34]. For example, we may apply the Gauss–Seidel method to (7.36) to obtain

$$\text{(7.38)}\quad \begin{aligned}\sum_{j\le i} a_{ij}\beta_j^{(r+1)} + \sum_{j>i} a_{ij}\beta_j^{(r)} - g_i(\beta_1^{(r+1)}, \ldots, \beta_i^{(r+1)}, \beta_{i+1}^{(r)}, \ldots, \beta_n^{(r)})\\ = 0, \qquad 1 \le i \le n.\end{aligned}$$

For each fixed i, $1 \le i \le n$, this equation is a nonlinear equation in the single unknown $\beta_i^{(r+1)}$ and has a unique solution, which can be obtained by Newton's method. Moreover, the cyclic determination of the $\beta_i^{(r+1)}$ is convergent; cf. [7.33] and [7.34].

Finally, we give some numerical results for a simple model problem. We consider

$$\text{(7.39)}\qquad -D^2u(x) = -\tfrac{1}{2}(u(x) + x + 1)^3, \qquad 0 < x < 1,$$

$$\text{(7.40)}\qquad u(0) = u(1) = 0,$$

which has the associated functional

$$\text{(7.41)}\qquad F[w] \equiv \int_0^1 [(Dw(x))^2 + \int_0^{w(x)} (\eta + x + 1)^3 d\eta]\, dx$$

and the unique solution $u(x) = 2(2 - x)^{-1} - x - 1$. The following computations for this problem were done by Dr. Robert Herbold; cf. [7.9]. All the partitions are uniform.

h	dim $L_0\Delta(h)$	$\|u - u_L\|_2$	dim $H_0\Delta(h)$	$\|u - u_H\|_2$	dim $S_0\Delta(h)$	$\|u - u_S\|_2$
$\frac{1}{3}$			6	$.19 \times 10^{-5}$		
$\frac{1}{4}$			8	$.75 \times 10^{-6}$	5	$.91 \times 10^{-6}$
$\frac{1}{5}$	4	$.15 \times 10^{-3}$	10	$.36 \times 10^{-6}$	6	$.47 \times 10^{-6}$
$\frac{1}{10}$	9	$.43 \times 10^{-4}$				
$\frac{1}{20}$	19	$.12 \times 10^{-4}$				

Though it is difficult to tell from these numerical results, we will prove in Section 7.5 that the RRG procedure is second-order accurate for $L_0(\Delta)$ and fourth-order accurate for $H_0(\Delta)$ and $S_0(\Delta)$ for all sufficiently smooth solutions.

7.4 SECOND-ORDER PROBLEMS IN THE PLANE

In this section, we consider the problem of approximating the solution of the second-order self-adjoint linear elliptic differential equation

$$
\begin{aligned}
-D_x[p(x, y)D_x u(x, y)] - D_y[r(x, y)D_y u(x, y)] + q(x, y)u(x, y) \\
= f(x, y), \qquad (x, y) \in \text{interior of } U,
\end{aligned}
\tag{7.42}
$$

subject to the Dirichlet boundary conditions

$$
u(x, y) = 0 \qquad \text{for all } (x, y) \in \text{boundary of } U,
\tag{7.43}
$$

where $p(x, y)$, $r(x, y)$, and $q(x, y) \in PC^{0,\infty}(U)$ are such that there exist positive constants γ and μ such that

$$
\begin{aligned}
\gamma \| u \|_D^2 &\equiv \gamma \int_0^1 \int_0^1 \{(D_x u)^2 + (D_y u)^2\}\, dxdy \\
&\leq \int_0^1 \int_0^1 \{p(x, y)(D_x u)^2 + r(x, y)(D_y u)^2 + q(x, y)u^2\}\, dxdy \\
&\leq \mu \| u \|_D^2
\end{aligned}
\tag{7.44}
$$

for all $u \in PC_0^{1,2}(U) \equiv \{\phi \in PC^{1,2}(U) \mid \phi(x, y) = 0 \text{ for all } (x, y) \in \text{boundary of } U\}$ and $f \in PC^{0,2}(U)$. We can extend the methods and results of this section to treat semilinear problems in the plane in the same way that we explicitly extended those of Section 7.2 to Section 7.3, for one-dimensional problems. The details are straightforward and are left to the reader. We say that u is a *generalized solution* (over $PC_0^{1,2}(U)$) of (7.42)–(7.43) if and only if $u \in PC_0^{1,2}(U)$ and

$$
\begin{aligned}
a(u, v) &\equiv \int_0^1 \int_0^1 \{p(x, y)D_x u D_x v + r(x, y)D_y u D_y v + q(x, y)uv\}\, dxdy \\
&= \int_0^1 \int_0^1 f(x, y)v\, dxdy
\end{aligned}
\tag{7.45}
$$

for all $v \in PC_0^{1,2}(U)$.

Integrating by parts, we can prove the following result.

THEOREM 7.10

If u is a classical solution of (7.42)–(7.43), then it is a generalized solution. Moreover, if the coefficients of the differential equation are sufficiently smooth, then every generalized solution is a classical solution.

We now state a variational characterization of generalized solutions. The proof is an exact analogue of the proof of Theorem 7.2.

THEOREM 7.11

The function $u(x, y)$ is a generalized solution of (7.42)–(7.43) if and only if $u(x, y)$ is the unique solution of the variational problem of finding u such that

$$F[u] = \inf_{w \in PC_0^{1,2}(U)} F[w] \equiv a(w, w) - 2(f, w)_2. \tag{7.47}$$

Following the treatment in Section 7.1, we let S be any finite-dimensional subspace of $PC_0^{1,2}(U)$ and $\{B_i(x, y)\}_{i=1}^n$ be a basis for S. The Rayleigh–Ritz procedure is to find an approximation to the generalized solution by determining an element $u_S \in S$ which minimizes $F[w]$ over S. The following result extends Theorem 7.3 to the two-dimensional case. The proof is analogous to that of Theorem 7.3.

THEOREM 7.12

There is a unique element

$$u_S \equiv \sum_{i=1}^{n} \beta_i^* B_i(x, y) \in S$$

which minimizes $F[w]$ over S. Moreover, $\boldsymbol{\beta}^*$ is the solution of the linear system with symmetric, positive definite matrix $A \equiv [a_{ij}] \equiv [a(B_i, B_j)]$.

If we define the Galerkin approximation $w_S(x, y) \equiv \sum_{i=1}^{n} \gamma_i B_i(x, y)$ to be determined by the solution of the linear equations

$$a(w_S, B_i) = (f, B_i)_2, \qquad 1 \leq i \leq n, \tag{7.48}$$

then we can see that the coefficients satisfy the exact same linear equations as $\boldsymbol{\beta}^*$ and hence $\boldsymbol{\gamma} = \boldsymbol{\beta}^*$, and we will refer to the Rayleigh–Ritz–Galerkin, RRG, approximation.

Now we turn our attention to the question of constructing suitable basis functions. As a basis for $L_0(\rho) \equiv \{l(x, y) \in L(\rho) \,|\, l(x, y) = 0$ for all $(x, y) \in$ boundary of $U\}$, we suggest the functions $\{l_i(x)l_j(y)\}_{i=1, j=1}^{N, M}$. This choice yields an RRG system with a sparse matrix having only nine nonzero diagonals. Moreover, the zero structure of the matrix is exactly the same as that of the least squares matrix for $L(\rho)$ described in Section 6.2, and we may use

Cholesky decomposition or successive overrelaxation to solve the linear system as in Section 6.2. It is possible to prove an analogue of Theorem 7.4 for $A_{L_0(\rho)}$; cf. [7.20].

As a basis for $H_0(\rho) \equiv \{h(x, y) \in H(\rho) \mid h(x, y) = 0$ for all $(x, y) \in$ boundary of $U\}$, we suggest the product basis functions

$$\{h_i(x)h_j(y)\}_{i=1,\, j=1}^{N,\, M} \cup \{\tilde{h}_i^1(x)h_j(y)\}_{i=0,\, j=1}^{N+1,\, M} \cup \{h_i(x)\tilde{h}_j^1(y)\}_{i=1,\, j=0}^{N,\, M+1} \cup \{\tilde{h}_i^1(x)\tilde{h}_j^1(y)\}_{i=0,\, j=0}^{N+1,\, M+1},$$

i.e., $H_0(\rho)$ has dimension $4(N + 1)(M + 1)$. This choice yields a sparse RRG matrix with thirty-six nonzero diagonals, as described in Section 6.2, and we may use either Cholesky decomposition or successive overrelaxation to solve the corresponding linear system.

As a basis for $S_0(\rho) \equiv \{s(x, y) \in S(\rho) \mid s(x, y) = 0$ for all $(x, y) \in$ boundary of $U\}$, we suggest the product basis functions $\{\tilde{s}_i(x)\tilde{s}_j(y)\}_{i=-2,\, j=-2}^{N-1,\, M-1}$. These functions yield a sparse RRG matrix, with forty-nine nonzero diagonals, which has essentially the same zero structure as the least squares matrix for $S(\rho)$ described in Section 6.2. Again we may use either Cholesky decomposition or successive overrelaxation to solve the RRG system.

Finally, we consider the two-dimensional problem

$$(7.49) \quad -D_x^2u(x, y) - D_y^2u(x, y) = -6xye^xe^y(xy + x + y - 3), \quad (x, y) \in \text{interior of } U,$$

$$(7.50) \quad u(x, y) = 0, \quad (x, y) \in \text{boundary of } U,$$

which has the associated functional

$$(7.51) \quad F[w] \equiv \int_0^1 \int_0^1 \{[(D_xw(x, y))^2 + (D_yw(x, y))^2] + 12[xye^xe^y(xy + x + y - 3)w(x, y)]\}\, dxdy$$

and the unique solution $u(x, y) = 3e^xe^y(x - x^2)(y - y^2)$. The following computations for this problem were also done by Dr. Robert Herbold; cf. [7.21]. All the partitions are uniform.

h & k	dim $L_0(\rho)$	$\lVert u - u_L \rVert_2$	dim $H_0(\rho)$	$\lVert u - u_H \rVert_2$	dim $S_0(\rho)$	$\lVert u - u_S \rVert_2$
$\frac{1}{3}$			36	$.92 \times 10^{-5}$	16	$.11 \times 10^{-4}$
$\frac{1}{4}$			64	$.32 \times 10^{-5}$	25	$.36 \times 10^{-5}$
$\frac{1}{5}$			100	$.14 \times 10^{-5}$	36	$.16 \times 10^{-5}$
$\frac{1}{6}$			144	$.71 \times 10^{-6}$	49	$.77 \times 10^{-6}$
$\frac{1}{7}$	36	$.31 \times 10^{-3}$			64	$.42 \times 10^{-6}$
$\frac{1}{8}$	49	$.25 \times 10^{-3}$				
$\frac{1}{9}$	64	$.20 \times 10^{-3}$				
$\frac{1}{10}$	81	$.16 \times 10^{-3}$				
$\frac{1}{11}$	100	$.14 \times 10^{-3}$				

These results indicate that for this particular problem the RRG procedure is second-order accurate for $L_0(\rho)$, and fourth-order accurate for $H_0(\rho)$ and $S_0(\rho)$. In Section 7.5, we will prove that this is true for all sufficiently smooth solutions.

7.5 ERROR ANALYSIS

In this section, we prove a priori error bounds for the Rayleigh–Ritz–Galerkin procedures introduced in Sections 7.2, 7.3, and 7.4. Throughout this section, we will assume that the generalized solution, u, exists. Our analysis is based upon the results of Chapters 2, 3, 4, and 6. We start with one-dimensional problems.

Before discussing error bounds there is one further characterization we want to introduce. From (7.3) it follows that $a(u, v)$ is an inner product on $PC_0^{1,2}(I)$ and we may show that u_S is the orthogonal projection of u onto S with respect to the inner product $a(u, v)$, i.e., $u = u_S + e_S$ where e_S is orthogonal to S or $a(e_S, B_i) = 0$ for all $1 \leq i \leq n$.

THEOREM 7.13

The RRG approximation, u_S, is the orthogonal projection with respect to the inner product $a(u, v)$ of u onto S, i.e., $a(u - u_S, B_i) = 0$ for all $1 \leq i \leq n$.

Proof. This result follows directly by subtracting (7.12) from (7.4) with $v \equiv B_i$. Q.E.D.

We proceed now to a discussion of a priori error bounds.

THEOREM 7.14

If u is the generalized solution of (7.1)–(7.2), S is any finite-dimensional subspace of $PC_0^{1,2}(I)$, and u_S is the RRG approximate to u, then

$$a(u - u_S, u - u_S) = \inf_{y \in S} a(u - y, u - y). \tag{7.52}$$

Proof. We will give two proofs of this fundamental result. First let $\{B_i\}_{i=1}^n$ be an orthonormal basis for S with respect to the inner product $a(u, v)$. Then, if $u_S = \sum_{i=1}^n \beta_i B_i$, $u = \sum_{i=1}^n \beta_i B_i + e_S$ where e_S is orthogonal to S. If $y = \sum_{i=1}^n \gamma_i B_i$ is any other element in S, we have

$$u = \sum_{i=1}^n \gamma_i B_i + \sum_{i=1}^n (\beta_i - \gamma_i) B_i + e_S$$

and

$$a(u - y, u - y) = \sum_{i=1}^n (\beta_i - \gamma_i)^2 + a(e_S, e_S),$$

which is minimized for $\beta_i = \gamma_i$, $1 \leq i \leq n$, i.e., for $y = u_S$.

For the second proof, we use (7.6). In fact, using (7.6) for $\eta = u_S$ and $\eta = y$, we have

$$\inf_{y \in S} a(u - y, u - y) \leq a(u - u_S, u - u_S) = F[u_S] - F[u]$$
$$\leq F[y] - F[u] = a(u - y, u - y).$$

Q.E.D.

We can now use this result in conjunction with the results of Chapters 2, 3, 4, and 6, giving a priori error bounds for interpolation mappings to obtain explicit a priori error bounds in the norm $\| D(u - u_S) \|_2$. Moreover, for a large class of problems, we can actually give sharp bounds for $\| u - u_S \|_2$. To this end, we say that (7.1)–(7.2) or $a(u, v)$ is *strongly coercive* if and only if there exists a positive constant, Γ, such that (7.1)–(7.2) has a generalized solution, $u \in PC^{2,2}(I) \cap PC_0^{1,2}(I)$, for all $f \in PC^{0,2}(I)$, and

(7.53) $$\| D^2 u \|_2 \leq \Gamma \| f \|_2, \qquad \text{for all } f \in PC^{0,2}(I).$$

If $p(x) \in PC^{1,\infty}(I)$, $p(x) \geq \delta > 0$ for all $x \in I$, and $q(x) \in PC^{0,\infty}(I)$, then (7.1)–(7.2) is strongly coercive and

$$\| D^2 u \|_2 \leq (\delta \gamma \pi)^{-1}(\| Dp \|_\infty + \pi^{-1} \| q \|_\infty + 1) \| f \|_2.$$

In fact, differentiating out (7.1), we have

(7.54) $$-D^2 u = \frac{1}{p(x)}[Dp(x) Du - q(x)u + f],$$

and hence

(7.55) $$\| D^2 u \|_2 \leq \left\| \frac{1}{p} \right\|_\infty [\| Dp \|_\infty \| Du \|_2 + \| q \|_\infty \| u \|_2 + \| f \|_2].$$

But

$$\gamma \| Du \|_2^2 \leq a(u, u) = (f, u)_2 \leq \| f \|_2 \| u \|_2 \leq \pi^{-1} \| f \|_2 \| Du \|_2$$

implies $\| Du \|_2 \leq (\gamma \pi)^{-1} \| f \|_2$ and

$$\| u \|_2 \leq \pi^{-1} \| Du \|_2 \leq \gamma^{-1} \pi^{-2} \| f \|_2.$$

Using these inequalities to bound the right-hand side of (7.55), we obtain the stated a priori bound.

THEOREM 7.15

If (7.1)–(7.2) is strongly coercive, then

(7.56) $$\| D(u - u_L) \|_2 \leq \gamma^{-1/2} \mu^{1/2} \pi^{-1} \Gamma \| f \|_2 h$$

and

$$\|u - u_L\|_2 \leq \gamma^{-1/2}\mu^{3/2}\Gamma^2\pi^{-2}\|f\|_2 h^2 \tag{7.57}$$

where u_L is the RRG approximation to u over

$$L_0(\Delta) \equiv \{l(x) \in L(\Delta) \mid l(0) = l(1) = 0\}.$$

Proof. From the result of Theorem 7.14 and inequality (7.3) we have

$$\begin{aligned} \gamma\|D(u - u_L)\|_2^2 &\leq a(u - u_L, u - u_L) \\ &\leq a(u - \vartheta_L u, u - \vartheta_L u) \\ &\leq \mu\|D(u - \vartheta_L u)\|_2^2, \end{aligned} \tag{7.58}$$

since $\vartheta_L u \in L_0(\Delta)$. The bound (7.56) follows by using the results of Theorem 2.5 to bound the right-hand side of (7.58) and by taking the square root of both sides of the resulting inequality.

To prove (7.57), we use a technique of Nitsche; cf. [7.28]. We let

$$\psi_L(x) \equiv (\|u - u_L\|_2)^{-1}(u(x) - u_L(x))$$

and consider the problem of finding ϕ_L such that

$$a(\phi_L, v) = (\psi_L, v)_2, \qquad \text{for all } v \in PC_0^{1,2}(I). \tag{7.59}$$

Since (7.1)–(7.2) is strongly coercive, the problem (7.59) has a unique solution, ϕ_L, and $\|D^2\phi_L\|_2 \leq \Gamma$. Moreover, $a(\phi_L, u - u_L) = \|u - u_L\|_2$.

Since $\vartheta_L\phi_L \in L_0(\Delta)$, we have by the definition of the RRG procedure,

$$a(\phi_L - \vartheta_L\phi_L, u - u_L) = a(\phi_L, u - u_L) = \|u - u_L\|_2,$$

and hence

$$\|u - u_L\|_2 \leq \mu\|D(\phi_L - \vartheta_L\phi_L)\|_2\|D(u - u_L)\|_2.$$

Using the inequalities (2.17) and (7.56) to bound the right-hand side of this inequality, we have

$$\begin{aligned} \|u - u_L\|_2 &\leq \mu(\pi^{-1}h\Gamma)(\gamma^{-1}\mu)^{1/2}\pi^{-1}\Gamma\|f\|_2 h \\ &= \gamma^{-1/2}\mu^{3/2}\Gamma^2\pi^{-2}\|f\|_2 h^2, \end{aligned}$$

which proves (7.57). Q.E.D.

In a similar way, we can prove the following two results.

THEOREM 7.16

If $u \in PC^{4,2}(I) \cap PC_0^{1,2}(I)$, then

$$\| D(u - u_H) \|_2 \leq \gamma^{-1/2} \mu^{1/2} \pi^{-3} h^3 \| D^4 u \|_2, \tag{7.60}$$

where u_H is the RRG approximation to u over

$$H_0(\Delta) \equiv \{h(x) \in H(\Delta) \mid h(0) = h(1) = 0\}.$$

Moreover, if in addition (7.1)–(7.2) is strongly coercive, then

$$\| u - u_H \|_2 \leq \gamma^{-1/2} \mu^{3/2} \Gamma \pi^{-4} h^4 \| D^4 u \|_2. \tag{7.61}$$

THEOREM 7.17

If $u \in PC^{4,2}(I) \cap PC_0^{1,2}(I)$, then

$$\| D(u - u_S) \|_2 \leq 2\gamma^{-1/2} \mu^{1/2} \pi^{-3} h^3 \| D^4 u \|_2, \tag{7.62}$$

where u_S is the RRG approximation to u over

$$S_0(\Delta) \equiv \{s(x) \in S(\Delta) \mid s(0) = s(1) = 0\}.$$

Moreover, if in addition (7.1)–(7.2) is strongly coercive, then

$$\| u - u_S \|_2 \leq 4\gamma^{-1/2} \mu^{3/2} \Gamma \pi^{-4} h^4 \| D^4 u \|_2. \tag{7.63}$$

Thus, we have shown that under suitable hypothesis the RRG approximation to u over $L_0(\Delta)$ is second-order accurate with respect to the L^2-norm, while the RRG approximations to u over $H_0(\Delta)$ and $S_0(\Delta)$ are both fourth-order accurate with respect to the L^2-norm. Roughly speaking, this means that asymptotically it is no more difficult to obtain a finite element RRG approximation to the solution of a linear two-point boundary value problem than it is to obtain a finite element least squares approximation given the solution.

We proceed now to discuss semilinear problems.

THEOREM 7.18

If u is the generalized solution of (7.19)–(7.20), S is any finite-dimensional subspace of $PC_0^{1,2}(I)$, and u_S is the RRG approximation to u over S, then

$$\begin{aligned} \| D(u - u_S) \|_2 &\leq \gamma^{-1}(1 - \lambda\Lambda^{-1})^{-1}(\mu + B\pi^{-2}) \inf_{y \in S} \| D(u - y) \|_2 \\ &\equiv \Omega \inf_{y \in S} \| D(u - y) \|_2. \end{aligned} \tag{7.64}$$

Proof. Since $a(u, B_i) = (f(u), B_i)_2$, $1 \leq i \leq n$, and

$$a(u_S, B_i) = (f(u_S), B_i)_2, \qquad 1 \leq i \leq n,$$

we have $a(u - u_S, B_i) = (f(u) - f(u_S), B_i)$, $1 \leq i \leq n$. But from the proof of Theorem 7.8 we have, for all $y \in S$,

$$\begin{aligned}
&\gamma(1 - \lambda\Lambda^{-1}) \| D(u - u_S) \|_2^2 \\
&\quad \leq a(u - u_S, u - u_S) - (f(u) - f(u_S), u - u_S)_2 \\
&\quad \leq a(u - u_S, u - y) - (f(u) - f(u_S), u - y)_2 \\
&\quad \leq \mu \| D(u - u_S) \|_2 \| D(u - y) \|_2 + B \| u - u_S \|_2 \| u - y \|_2 \\
&\quad \leq \mu \| D(u - u_S) \|_2 \| D(u - y) \|_2 + B\pi^{-2} \| D(u - u_S) \|_2 \| D(u - y) \|_2,
\end{aligned}$$

and hence

$$\| D(u - u_S) \|_2 \leq \gamma^{-1}(1 - \lambda\Lambda^{-1})^{-1}(\mu + B\pi^{-2}) \| D(u - y) \|_2. \tag{7.65}$$

Q.E.D.

We can now use this result in conjunction with the appropriate results from Chapters 2, 3, 4, and 6 to obtain explicit a priori error bounds.

THEOREM 7.19

If $u \in PC^{2,2}(I) \cap PC_0^{1,2}(I)$, then

$$\| D(u - u_L) \|_2 \leq \Omega\pi^{-1} h \| D^2 u \|_2, \tag{7.66}$$

where Ω is the constant defined in (7.64) and u_L denotes the RRG approximation to u over $L_0(\Delta)$. Moreover, if in addition $a(u, v)$ is strongly coercive, then

$$\| u - u_L \|_2 \leq \Omega^2 \pi^{-2} \Gamma(\mu + B\pi^{-2}) h^2 \| D^2 u \|_2. \tag{7.67}$$

Proof. The bound (7.66) follows by using the results of Theorem 2.5 to bound the right-hand side of (7.64) of Theorem 7.18. To prove (7.67), let

$$\psi_L(x) \equiv (\| u - u_L \|_2)^{-1}(u(x) - u_L(x))$$

and let $\phi_L(x)$ be the unique solution of the *linear* problem

$$\begin{aligned}
b(v, \phi_L) &\equiv a(v, \phi_L) + \left(\frac{\partial f}{\partial u}[\theta u + (1 - \theta)u_L]v, \phi_L\right)_2 \\
&= (\psi_L, v)_2,
\end{aligned} \tag{7.68}$$

where $0 < \theta < 1$, for all $v \in PC_0^{1,2}(I)$.

Then

$$\gamma(1 - \lambda\Lambda^{-1}) \| Dv \|_2^2 \leq (1 - \lambda\Lambda^{-1})a(v, v) \leq b(v, v) \leq \mu \| Dv \|_2^2 + B \| v \|_2^2 \leq (\mu + B\pi^{-2}) \| Dv \|_2^2, \tag{7.69}$$

for all $v \in PC_0^{1,2}(I)$, and

$$\gamma(1 - \lambda\Lambda^{-1}) \| D\phi_L \|_2^2 \leq b(\phi_L, \phi_L) = (\psi_L, \phi_L)_2 \leq \| \psi_L \|_2 \| \phi_L \|_2 \leq \pi^{-1} \| D\phi_L \|_2. \tag{7.70}$$

From the strong coerciveness of $a(u, v)$ and (7.70) we have

$$\| D^2\phi_L \|_2 \leq \Gamma[B\gamma^{-1}(1 - \lambda\Lambda^{-1})^{-1}\pi^{-2} + 1] \leq \Gamma\Omega. \tag{7.71}$$

Moreover, from (7.68) we have that for all $l \in L(\Delta)$,

$$\begin{aligned} \| u - u_L \|_2 &= a(u - u_L, \phi_L - l) + \left(\frac{\partial f}{\partial u}(u - u_L), \phi_L - l\right)_2 \\ &\leq \mu \| D(u - u_L) \|_2 \| D(\phi_L - l) \|_2 \\ &\quad + B\pi^{-2} \| D(u - u_L) \|_2 \| D(\phi_L - l) \|_2 \\ &\leq (\mu + B\pi^{-2}) \| D(u - u_L) \|_2 \| D(\phi_L - y) \|_2. \end{aligned} \tag{7.72}$$

Choosing $y \equiv \vartheta_L u$, we may use (7.66) and (2.17) to bound the right-hand side of (7.72) and obtain (7.67). Q.E.D.

In a similar way, we can prove the following two results.

THEOREM 7.20

If $u \in PC^{4,2}(I) \cap PC_0^{1,2}(I)$, then

$$\| D(u - u_H) \|_2 \leq \Omega\pi^{-3}h^3 \| D^4u \|_2,$$

where Ω is the constant defined in (7.64) and u_H denotes the RRG approximation to u over $H_0(\Delta)$. Moreover, if in addition $a(u, v)$ is strongly coercive, then

$$\| u - u_H \|_2 \leq \Omega^2\pi^{-4}\Gamma(\mu + B\pi^{-2})h^4 \| D^4u \|_2. \tag{7.74}$$

THEOREM 7.21

If $u \in PC^{4,2}(I) \cap PC_0^{1,2}(I)$, then

$$\| D(u - u_S) \|_2 \leq 2\Omega\pi^{-3}h^3 \| D^4u \|_2, \tag{7.75}$$

where Ω is the constant defined in (7.64) and u_S denotes the RRG approximation to u over $S_0(\Delta)$. Moreover if in addition $a(u, v)$ is coercive, then

$$\|u - u_S\|_2 \leq 4\Omega^2\pi^{-4}\Gamma(\mu + B\pi^{-2})h^4 \|D^4u\|_2. \tag{7.76}$$

We turn now to a discussion of the two-dimensional problem. Analogues of Theorems 7.13 and 7.14 hold.

THEOREM 7.22

If u is the generalized solution of (7.59)–(7.60), S is any finite-dimensional subspace of $PC_0^{1,2}(U)$, and u_S is the *RRG* approximate to u, then

$$a(u - u_S, u - u_S) = \inf_{y \in S} a(u - y, u - y). \tag{7.77}$$

We can now use this result in conjunction with the results of Chapters 2, 3, 4, and 6 to obtain a priori error bounds in the norm $\|u - u_S\|_D$. Moreover, for a large class of problems, we can actually give sharp error bounds for $\|u - u_S\|_2$. To this end, we say that (7.42)–(7.43) or $a(u, v)$ is *strongly coercive* if and only if (7.42)–(7.43) has a generalized solution for all $f \in PC^{0,2}(U)$,

$$u \in PC^{2,2}(U) \cap PC_0^{1,2}(U),$$

and

$$\|D_x^k D_y^j u\|_2 \leq \Gamma \|f\|_2, \qquad \text{for all } 0 \leq k + j \leq 2. \tag{7.78}$$

Birman and Skvortsov have shown that if $p(x, y)$ and $r(x, y) \in C^1(U)$, $q(x, y) \in C(U)$, and $a(u, v)$ satisfies (7.44), then $a(u, v)$ is strongly coercive; cf. [7.5].

THEOREM 7.23

If (7.42)–(7.43) is strongly coercive, then

$$\|u - u_L\|_D \leq 6\gamma^{-1/2}\mu^{1/2}\Gamma\pi^{-1} \|f\|_2\bar{\rho}, \tag{7.79}$$

where u_L is the RRG approximation to $\mathbf{u}$ over

$$L_0(\rho) \equiv \{l(x, y) \in L(\rho) \mid l(x, y) = 0 \text{ for all } (x, y) \in \text{boundary of } U\},$$

and

$$\|u - u_L\|_2 \leq 36\gamma^{-1/2}\mu^{3/2}\Gamma^2\pi^{-2} \|f\|_2\bar{\rho}^2. \tag{7.80}$$

Proof. From the result of Theorem 7.22 and inequality (7.77) we have

$$\begin{aligned} \gamma\|u - u_L\|_D^2 &\leq a(u - u_L, u - u_L) \leq a(u - \vartheta_L u, u - \vartheta_L u) \\ &\leq \mu\|u - \vartheta_L u\|_D^2, \end{aligned} \tag{7.81}$$

since $\vartheta_L u \in L_0(\rho)$. The bound (7.79) follows by taking the square root of both sides of (7.81) and using the results of Theorem 2.7 to bound the right-hand side of the resulting inequality.

To prove (7.80), we can follow the proof of Theorem 7.15 almost exactly to obtain the inequality

$$\|u - u_L\|_2 \leq \mu \|\phi_L - \vartheta_L \phi_L\|_D \|u - u_L\|_D, \tag{7.82}$$

where ϕ_L is the solution of the two-dimensional analogue of (7.59). Using the results of Theorem 2.7 and inequality (7.79) to bound the right-hand side of (7.82), we obtain (7.80). Q.E.D.

In a similar way, we can prove the following two results.

THEOREM 7.24

If $u \in PC^{4,2}(U) \cap PC_0^{1,2}(U)$, then

$$\|u - u_H\|_D \leq \gamma^{-1/2}\mu^{1/2}\pi^{-3}(1 + \pi^{-1}2^{1/2}9)\bar{\rho}^3 \|u\|_{4,2}, \tag{7.83}$$

where $\|u\|_{4,2} \equiv \sum_{k+j=4} \|D_x^k D_y^j u\|_2$ and u_H is the RRG approximation to u over

$$H_0(\rho) \equiv \{h(x, y) \in H(\rho) \mid h(x, y) = 0 \text{ for all } (x, y) \in \text{boundary of } U\}.$$

Moreover, if in addition (7.42)–(7.43) is strongly coercive, then

$$\begin{aligned}\|u - u_H\|_2 \leq {} & 2\Gamma\gamma^{-1/2}\mu^{3/2}\pi^{-4}(1 + \pi^{-1}2\sqrt{15}) \\ & \times (1 + 2\sqrt{90} + 2\sqrt{1605}\underline{\rho}^{-1}\bar{\rho})\bar{\rho}^4 \|u\|_{4,2}.\end{aligned} \tag{7.84}$$

Proof. In analogy to the preceding proof, we have

$$\|u - u_H\|_D \leq \gamma^{-1/2}\mu^{1/2} \|u - \vartheta_H u\|_D, \tag{7.85}$$

and (7.83) follows by using the results of Theorem 3.10 to bound the right-hand side of (7.85).

To prove (7.84) we can follow the proof of Theorem 7.16 almost exactly to obtain the inequality

$$\|u - u_H\|_2 \leq \mu \|\phi_H - P_{H(\rho)}\phi_H\|_D \|u - u_H\|_D, \tag{7.86}$$

where ϕ_H is the solution of the two-dimensional analogue of (7.59). Using the results of Exercise (6.5) and (7.83) to bound the right-hand side of (7.86), we obtain (7.84). Q.E.D.

THEOREM 7.25

If $u \in PC^{4,2}(U) \cap PC_0^{1,2}(U)$, then

$$\text{(7.87)} \quad \| u - u_S \|_D \leq 2\gamma^{-1/2}\mu^{1/2}\pi^{-3}(1 + 2\sqrt{90 + 2\sqrt{1605}}\underline{\rho}^{-1}\bar{\rho})\bar{\rho}^3 \| u \|_{4,2},$$

where u_S is the RRG approximation to u over

$$S_0(\rho) \equiv \{s(x, y) \in S(\rho) \,|\, s(x, y) = 0 \text{ for all } (x, y) \in \text{boundary of } U\}.$$

Moreover, if in addition (7.42)–(7.43) is strongly coercive, then

$$\text{(7.88)} \quad \| u - u_S \|_2 \leq 4\Gamma\gamma^{-1/2}\mu^{3/2}\pi^{-4}(1 + 2\sqrt{90 + 2\sqrt{1605}}\underline{\rho}^{-1}\bar{\rho})^2\bar{\rho}^4 \| u \|_{4,2}.$$

In general we would not expect $u \in PC^{4,2}(U)$ because of singularities of derivatives of u at the corners of R. However, it is possible to augment $H_0(\rho)$ and $S_0(\rho)$ with appropriate "singular" basis functions so that we rigorously obtain results which are essentially the same as those of Theorems 7.24 and 7.25 with the augmented spaces. See [7.17] for the details.

EXERCISES FOR CHAPTER 7

(7.1) Prove analogues of the results of Sections 7.2 and 7.3 for the spaces

$$S_0(2m - 1, \Delta, z) \equiv \{\phi \in S(2m - 1, \Delta, z) \,|\, \phi(0) = \phi(1) = 0\}$$

defined in Exercise (4.12) (cf. [7.39]). Prove analogues of the results of Section 7.4 for the spaces

$$S_0(2m - 1, \rho, z) \equiv S_0(2m - 1, \Delta, z) \otimes S_0(2m - 1, \Delta_y, z)$$

(cf. [7.39]).

(7.2) Prove analogues of the results of Section 7.2 for the linear two-point boundary value problem of order $2n$, i.e., find $u \in PC^{n,2}(I)$ such that

$$\sum_{i=0}^{n} (-1)^i D^i[p_i(x)D^i u(x)] = f(x), \qquad 0 < x < 1,$$

$$D^i u(0) = D^i u(1) = 0, \qquad 0 \leq i \leq n - 1,$$

where $p_i(x) \in PC^{0,\infty}(I)$, $0 \leq i \leq n$, and

$$\gamma \| D^n u \|_2^2 \leq \int_0^1 \sum_{i=0}^{n} p_i(x)[D^i u]^2 \, dx \leq \mu \| D^n u \|_2^2$$

for all

$$u \in PC_0^{n,2}(I) \equiv \{\phi \in PC^{n,2}(I) \,|\, D^i\phi(0) = D^i\phi(1) = 0, \ 0 \leq i \leq n - 1\}.$$

(Hint: Use the functional

$$F[w] \equiv \int_0^1 \sum_{i=0}^n p_i(x)[D^i w]^2 \, dx - 2\int_0^1 f(x) w \, dx;$$

cf. [7.9] and [7.39]).

(7.3) Prove analogues of the results of Section 7.3 for the nonlinear two-point boundary value problem of order $2n$, i.e., find $u \in PC^{n,2}(I)$ such that

$$\sum_{i=0}^n (-1)^i D^i[p_i(x) D^i u(x)] = f(x, u), \qquad 0 < x < 1,$$

$$D^i u(0) = D^i u(1) = 0, \qquad 0 \leq i \leq n-1,$$

where the coefficients $p_i(x)$, $0 \leq i \leq n$, satisfy the hypotheses of Exercise (7.2),

$$f(x, u), \frac{\partial f}{\partial u}(x, u) \in C([0, 1] \times (-\infty, \infty)),$$

$$\left|\frac{\partial f}{\partial u}(x, u)\right| \leq B \text{ for all } (x, u) \in [0, 1] \times (-\infty, \infty), \text{ and}$$

$$\frac{\partial f}{\partial u}(x, u) \leq \lambda < \Lambda \equiv \inf_{w \in PC_0^{n,2}(I)} \int_0^1 \frac{\sum_{i=0}^n p_i(x)[D^i w(x)]^2 \, dx}{\int_0^1 [w]^2 \, dx}$$

(Hint: Use the functional

$$F[w] \equiv \int_0^1 \left\{\sum_{i=0}^n p_i(x)[D^i w]^2 \, dx - 2\int_0^{w(x)} f(x, t) \, dt\right\} dx;$$

cf. [7.9] and [7.39].)

(7.4) Follow the example of Exercises (7.2) and (7.3) and generalize the results of Section 7.4 to problems of order $2n$ in the square (cf. [7.4] and [7.39]).

(7.5) In many nonlinear two-point boundary value problems, we have all the hypotheses of Section 7.3 except for the boundedness of $f(x, u)$ as a function of u. Show that if under these hypotheses we can still prove an a priori bound, $\| u \|_\infty \leq B$, for any solution of (7.25)–(7.26) then we may apply all the results of Section 7.3 to the equivalent problem

$$-D[p(x) D u_B] + q(x) u_B = f(x, \xi_B(u_B)), \; 0 < x < 1,$$

$$u_B(0) = u_B(1) = 0,$$

where

$$\xi_B(u_B) \equiv \begin{cases} B + 1 - e^{B - u_B}, & B < u_B, \\ u_B, & |u_B| \leq B, \\ -B - 1 + e^{B + u_B}, & u_B < -B, \end{cases}$$

i.e., the modified problem satisfies all the hypotheses of Section 7.3; if u_B is a solution of the modified problem, then $\| u_B \|_\infty \leq B$; and if u_B exists, then $u_B = u$ (cf. [7.12]).

(7.6) Show that for the problem $-D^2u = f(x)$, $0 < x < 1$, $u(0) = u(1) = 0$, $u_L = \vartheta_{L(\Delta)}u$ and hence u_L is "infinitely accurate" at the mesh points and if $f \in PC^{0,\infty}(I)$, then

$$\|u - u_L\|_\infty \leq \tfrac{1}{8}h^2\|f\|_\infty.$$

Moreover, show that if $f \in PC^{2,\infty}(I)$, then there exists a positive constant, K, such that

$$\|u - u_H\|_\infty \leq Kh^4 \|D^2 f\|_\infty$$

and

$$\|u - u_S\|_\infty \leq Kh^4 \|D^2 f\|_\infty.$$

(cf. [7.23], [7.30], and [7.36] for more general results on the uniform convergence of RRG approximations.)

(7.7) Consider the nonselfadjoint problem of finding $u(x)$ such that

$$-D^2u(x) + p(x)Du(x) + q(x)u(x) = f(x), \qquad 0 < x < 1,$$

and $u(0) = u(1) = 0$, where we assume that $p(x) \in C^1(I)$, $q(x) \in C^0(I)$, $f(x) \in PC^{0,2}(I)$, and there exists a positive constant γ such that

$$\gamma\|Dw\|_2^2 \leq \int_0^1 [(Dw)^2 + (q(x) - \tfrac{1}{2}Dp(x))(w)^2]\,dx$$

for all $w \in PC_0^{1,2}(I)$. We say that this problem has a generalized solution, u, if and only if

$$a(u, v) \equiv \int_0^1 [DuDv + p(x)vDu + q(x)uv]\,dx$$
$$= \int_0^1 f(x)v\,dx \equiv (f, v)_2$$

for all $v \in PC_0^{1,2}(I)$. Show that this problem has at most one generalized solution. Let S be any n-dimensional subspace of $PC_0^{1,2}(I)$ with basis $\{B_i(x)\}_{i=1}^n$. If the generalized solution exists, we define the Galerkin approximate, u_S, as the solution of

$$a(u_S, B_i) = (f, B_i)_2, \qquad 1 \leq i \leq n.$$

Show that u_S is well-defined and that

$$\|D(u - u_S)\|_2 \leq [1 + \gamma^{-1}(1 + P\pi^{-1} + Q\pi^{-2})] \inf_{y \in S} \|D(u - y)\|_2,$$

where

$$P \equiv \max_{x \in I}|p(x)| \quad \text{and} \quad Q \equiv \max_{x \in I}|q(x)|.$$

Prove analogues of inequalities (7.57), (7.61), and (7.63) for this problem.

REFERENCES FOR CHAPTER 7

[7.1] AUBIN, J.-P., Behavior of the error of the approximate solutions of boundary value problems for linear elliptic operators by Galerkin's and finite difference methods. *Annali della Scoula Normale di Pisa* **21**, 599–637 (1967).

[7.2] AUBIN, J.-P., and H. G. BURCHARD, Some aspects of the method of the hypercircle applied to elliptic variational problems. *Numerical Solution of Partial Differential Equations—II* (B. Hubbard, ed.) 1–68, Academic Press, New York (1971).

[7.3] BABUŠKA, Ivo, The finite element method for elliptic differential equations. *Numerical Solution of Partial Differential Equations—II* (B. Hubbard, ed.) 69–106, Academic Press, New York (1971).

[7.4] BIRKHOFF, G., M. H. SCHULTZ, and R. S. VARGA, Piecewise Hermite interpolation in one and two variables with applications to partial differential equations. *Numer. Math.* **11**, 232–256 (1968).

[7.5] BIRMAN, M. S. and G. E. SKVORTSOV, On the summability of the highest-order derivatives of the solution of the Dirichlet problem in a domain with piecewise smooth boundary. *Izv. Vyssh. Nchebn. Zaved Matematika* **30**, 12–21 (1962).

[7.6] BRAMBLE, J. H., and A. H. SCHATZ, On the numerical solution of elliptic boundary value problems by least squares approximation of the data. *Numerical Solution of Partial Differential Equations—II* (B. Hubbard, ed.) 107–132, Academic Press, New York (1971).

[7.7] BROWDER, F. E., Existence and uniqueness theorems for solutions of nonlinear boundary value problems. *Proc. Amer. Math. Soc. Symposia in Appl. Math.* **17**, 24–49 (1965).

[7.8] CÉA, J., Approximation variationnelle des problèmes aux limites. *Ann. Inst. Fourier* (*Grenoble*) **14**, 345–444 (1964).

[7.9] CIARLET, P. G., M. H. SCHULTZ, and R. S. VARGA, Numerical methods of high-order accuracy for nonlinear boundary value problems. I: One-dimensional problems. *Numer. Math.* **9**, 394–430 (1967).

[7.10] CIARLET, P. G., M. H. SCHULTZ, and R. S. VARGA, Numerical methods of high-order accuracy for nonlinear boundary value problems. II: Nonlinear boundary conditions. *Numer. Math.* **11**, 331–345 (1968).

[7.11] CIARLET, P. G., M. H. SCHULTZ, and R. S. VARGA, Numerical methods of high-order accuracy for nonlinear boundary value problems. IV: Periodic boundary conditions. *Numer. Math.* **12**, 266–279 (1968).

[7.12] CIARLET, P. G., M. H. SCHULTZ, and R. S. VARGA, Numerical methods of high-order accuracy for nonlinear boundary value problems. V: Monotone operators. *Numer. Math.* **13**, 51–77. (1969).

[7.13] CIARLET, P. G., F. NATTERER, and R. S. VARGA, Numerical methods of high-order accuracy for singular nonlinear boundary value problems. *Numer. Math.* **15**, 87–99 (1970).

[7.14] COLLATZ, L., *The Numerical Treatment of Differential Equations*, 3rd ed. Springer-Verlag, Berlin (1960).

[7.15] COURANT, R., Variational methods for the solution of problems of equilibrium and vibrations. *Bull. Amer. Math. Soc.* **49**, 1–23 (1943).

[7.16] COURANT, R., D. HILBERT, *Methods of Mathematical Physics.* Interscience, New York (1962).

[7.17] FIX, G., Higher-order Rayleigh–Ritz approximations. *J Math. and Mech.* **18**, 645–658 (1969).

[7.18] FRIEDRICHS, K. O., and H. B. KELLER, A finite difference scheme for generalized Neumann problems. *Numerical Solution of Partial Differential Equations* (J. Bramble, ed.) 1–19, Academic Press, New York (1966).

[7.19] GEORGE, J. A., *Computer implementation of the finite element method.* Stanford Computer Science Research Report STAN–CS–71–208 (1971).

[7.20] GILLON, A., and M. H. SCHULTZ, *On the conditioning of matrices arising in the finite element method.* Yale Computer Science Research Report.

[7.21] HERBOLD, R. J., *Consistent quadrature schemes for the numerical solution of boundary value problems by variational techniques.* Ph.D. Dissertation, Case Western Reserve University (June, 1968).

[7.22] HERBOLD, R. J., M. H. SCHULTZ, and R. S. VARGA, Quadrature schemes for the numerical solution of boundary value problems. *Aequationes Mathematicae* **3**, 247–270 (1970).

[7.23] HULME, B. L., Interpolation by Ritz approximation. *J. Math. and Mech.* **18**, 337–341 (1968).

[7.24] ISAACSON, E., and H. B. KELLER, *Analysis of Numerical Methods.* John Wiley & Sons, Inc., New York (1966).

[7.25] KANTOROVICH, J. V., and V. I. KRYLOV, *Approximate Methods of Higher Analysis.* Interscience, New York (1958).

[7.26] MIKHLIN, S. G., *Variational Methods in Mathematical Physics.* Macmillan, New York (1964).

[7.27] MIKHLIN, S. G., and K. L. SMOLITSKIY, *Approximate Methods for Solution of Differential and Integral Equations*, American Elsevier Publishing Co., Inc. (1967).

[7.28] NITSCHE, J., Ein Kriterium für die Quasi-Optimalität des Ritzschen Verfahrens. *Numer. Math.* **11**, 346–348 (1968).

[7.29] ORTEGA, J. M., and M. L. ROCKOFF, Nonlinear difference equations and Gauss–Seidel type iterative methods. *SIAM J. Numer. Anal.* **3**, 497–513 (1966).

[7.30] PERRIN, F. M., H. S. PRICE, and R. S. VARGA, On higher-order methods for nonlinear two-point boundary value problems. *Numer. Math.* **13**, 180–198 (1969).

[7.31] PETRYSHYN, W. V., Direct and iterative methods for the solution of linear operator equations in Hilbert space. *Trans. Amer. Math. Soc.* **105**, 136–175 (1962).

[7.32] PETRYSHYN, W. V., On nonlinear *P*-compact operators in Banach space with applications to constructive fixed-point theorems. *J. Math. Anal. and Appl.* **15**, 228–242 (1966).

[7.33] SCHECTER, S., Iteration methods for nonlinear problems. *Trans. Amer. Math. Soc.* **104**, 179–189 (1962).

[7.34] SCHECTER, S., Relaxation methods for convex problems. *SIAM J. Numer. Anal.* **5**, 601–612 (1968).

[7.35] SCHULTZ, M. H., Error bounds for the Rayleigh–Ritz–Galerkin method. *J. Math. Anal. and Appl.* **27**, 524–533 (1969).

[7.36] SCHULTZ, M. H., Elliptic spline functions and the Rayleigh–Ritz–Galerkin method. *Math. Comp.* **24**, 65–80 (1970).

[7.37] SCHULTZ, M. H., Rayleigh–Ritz–Galerkin methods for multi-dimensional problems. *SIAM J. Numer. Anal.* **6**, 523–538 (1969).

[7.38] SCHULTZ, M. H., Multivariate spline functions and elliptic problems. *Approximations with Special Emphasis on Spline Functions* (I. J. Schoenberg, ed.), 279–347, Academic Press, New York (1969).

[7.39] SCHULTZ, M. H., L^2-error bounds for the Rayleigh–Ritz–Galerkin method. *SIAM J. Numer. Anal.* **8**, 737–748 (1971).

[7.40] SCHULTZ, M. H., *Quadrature–Galerkin approximations to solutions of elliptic differential equations.* Yale Computer Science Research Report 71–8 (1971).

[7.41] SCHULTZ, M. H., *Quadrature methods for implementing the Rayleigh–Ritz–Galerkin method.* Yale Computer Science Research Report.

[7.42] STRANG, G., The finite element method and approximation theory. *Numerical Solution of Partial Differential Equations—II* (B. Hubbard, ed.) 547–584, Academic Press, New York (1971).

[7.43] VARGA, R. S., *Matrix Iterative Analysis.* Prentice-Hall, Inc., Englewood Cliffs, N. J. (1962).

[7.44] ZIENKIEWICZ, O. C., *The Finite Element Method in Structural and Continuum Mechanics.* McGraw-Hill, London, (1967).

[7.45] ZLÁMAL, M., On the finite element method. *Numer. Math.* **12**, 394–409 (1968).

[7.46] ZLÁMAL, M., On some finite element procedures for solving second order boundary value problems. *Numer. Math.* **14**, 42–48 (1969).

[7.47] ZLÁMAL, M., A finite element procedure of the second order accuracy. *Numer. Math.* **14**, 394–402 (1970).

8 THE RAYLEIGH-RITZ-GALERKIN PROCEDURE FOR EIGENVALUE PROBLEMS

8.1 INTRODUCTION

In this chapter, we consider the problem of finding those real numbers, λ, such that the problem

$$(8.1) \qquad -D[p(x)Du] + q(x)u = \lambda r(x)u, \qquad 0 < x < 1,$$

$$(8.2) \qquad u(0) = u(1) = 0,$$

has a nontrivial solution. A value of λ for which a nontrivial solution of (8.1)–(8.2) exists is called an *eigenvalue* and the nontrivial solution $u(x)$ is called an *eigenfunction*.

The results of this chapter extend to higher-order problems, problems with more general boundary conditions, and problems in more than one independent variable; cf. [8.1], [8.3], [8.10], and [8.12] for further details.

8.2 ONE-DIMENSIONAL PROBLEMS

We make the assumption that $p(x) \in C^1(I)$, that $q(x)$ and $r(x) \in C^0(I)$, and that there exist positive constants α, γ, and μ such that

$$(8.3) \qquad \begin{aligned} \alpha\gamma \| w \|_2^2 \leq \alpha b(w, w) \equiv \alpha \int_0^1 r(x)[w]^2 \, dx \leq a(w, w) \\ \equiv \int_0^1 \{p(x)[Dw]^2 + q(x)[w]^2\} \, dx \leq \mu \| Dw \|_2^2, \end{aligned}$$

for all $w \in PC_0^{1,2}(I)$. It is well-known that solutions $\{\lambda, u(x)\}$ of (8.1)–(8.2) can be characterized in terms of the minimum values and minimizing functions of the *Rayleigh-quotient*:

$$R[w] \equiv \frac{a(w, w)}{b(w, w)} \equiv \frac{a[w]}{b[w]}, \tag{8.4}$$

$w \in PC_0^{1,2}(I)$, $w \not\equiv 0$; cf. [8.4], [8.5], [8.6], and [8.9]. We now state the results of Brauer (cf. [8.2]) and Kamke (cf. [8.7] and [8.8]) for this problem.

THEOREM 8.1

The problem (8.1)–(8.2) has countably many eigenvalues which are real, have no finite limit point, and can be arranged as

$$0 < \lambda_1 \leq \lambda_2 \leq \dots . \tag{8.5}$$

There is a corresponding sequence of eigenfunctions $\{u_j(x)\}_{j=1}^{\infty}$, where $u_j \in C^2(I)$ and

$$-D[p(x)Du_j] + q(x)u_j = \lambda_j r(x)u_j, \qquad 0 < x < 1,$$

for all $j \geq 1$, and these eigenfunctions can be normalized so that

$$a(u_i, u_j) = \lambda_i \delta_{ij}, \qquad \text{for all } i, j = 1, 2, \dots, \tag{8.6}$$

and

$$b(u_i, u_j) = \delta_{ij}, \qquad \text{for all } i, j = 1, 2, \dots . \tag{8.7}$$

Moreover, each eigenvalue, λ_j, $j \geq 1$, can be characterized as

$$\lambda_j = \begin{cases} \inf\{R[w] \mid w \in PC_0^{1,2}(I) \text{ such that } b(w, u_k) = 0,\ 1 \leq k < j\}, \\ R[u_j], \\ \min\left\{\max\limits_{c_1,\dots,c_j} R\left[\sum\limits_{i=1}^{j} c_i v_i\right] \,\middle|\, v_1(x), \dots, v_j(x) \in PC_0^{1,2}(I) \\ \hfill \text{linearly independent}\right\}. \end{cases} \tag{8.8}$$

As in the previous chapters, let S be a finite-dimensional subspace of $PC_0^{1,2}(I)$ spanned by the basis functions $\{B_i(x)\}_{i=1}^n$. The Rayleigh–Ritz procedure consists of looking for the minimizing points of $R[w]$ over S. The restriction of R to S can be viewed as

$$\begin{aligned} R\left[\sum_{i=1}^n \beta_i B_i(x)\right] &= \frac{a\left(\sum_{i=1}^n \beta_i B_i, \sum_{i=1}^n \beta_i B_i\right)}{b\left(\sum_{i=1}^n \beta_i B_i, \sum_{i=1}^n \beta_i B_i\right)} \\ &= \left(a\left[\sum_{i=1}^n \beta_i B_i\right]\right)\left(b\left[\sum_{i=1}^n \beta_i B_i\right]\right)^{-1}. \end{aligned}$$

To find the minimum values of $R[\boldsymbol{\beta}]$, we write using the calculus

$$(8.9) \qquad \frac{\partial a\left[\sum_{i=1}^{n} \beta_i B_i\right]}{\partial \beta_i} = \lambda \frac{\partial b\left[\sum_{i=1}^{n} \beta_i B_i\right]}{\partial \beta_i}, \qquad 1 \le i \le n,$$

which yields the matrix eigenvalue problem,

$$(8.10) \qquad A\boldsymbol{\beta} = \lambda B\boldsymbol{\beta},$$

where the $n \times n$ matrices A and B have their entries given by

$$(8.11) \qquad A \equiv [a_{ij}] \equiv [a(B_i, B_j)]$$

and

$$(8.12) \qquad B \equiv [b_{ij}] \equiv [b(B_i, B_j)].$$

The matrices A and B are clearly symmetric and positive definite. Thus the matrix eigenvalue problem (8.10) has n positive eigenvalues $0 < \tilde{\lambda}_1 \le \tilde{\lambda}_2 \le \ldots \le \tilde{\lambda}_n$ and corresponding linearly independent eigenvectors $\tilde{\boldsymbol{\beta}}_1, \ldots, \tilde{\boldsymbol{\beta}}_n$. To each eigenvector $\tilde{\boldsymbol{\beta}}_j$, $1 \le j \le n$, we associate the function

$$(8.13) \qquad \tilde{u}_j(x) \equiv \sum_{i=1}^{n} \tilde{\beta}_{i,j} B_i(x),$$

where $\tilde{\beta}_{i,j}$ is the i-th component of $\tilde{\boldsymbol{\beta}}_j$, and henceforth we will call $\tilde{\lambda}_j$ an *approximate eigenvalue* and $\tilde{u}_j(x)$ an *approximate eigenfunction* for (8.1)–(8.2). Clearly, we have the following characterizations, which are analogues of (8.8).

$$\tilde{\lambda}_j = \begin{cases} R[\tilde{u}_j] \\ \inf\{R[w] \mid w \in S \text{ and } b(w, \tilde{u}_k) = 0,\ 1 \le k < j\} \\ \min\left\{\max\limits_{c_1,\ldots,c_j} R\left[\sum_{i=1}^{j} c_i v_i\right] \middle|\ v_1(x), \ldots, v_j(x) \in S \text{ linearly independent}\right\}. \end{cases}$$

The Galerkin procedure is to find an approximation $\{\hat{\lambda}_j, \hat{u}_j \in S\}$ by solving the finite-dimensional problem

$$a(\hat{u}_j, B_i) = \hat{\lambda}_j \int_0^1 r(x)\hat{u}_j B_i \, dx,$$

for all $1 \le i \le n$. If we expand $\hat{u}_j$ in terms of the basis functions, then we can see that the coefficients satisfy the matrix eigenvalue problem (8.10). Thus, for a problem of this form the Rayleigh–Ritz and Galerkin approximations are identical and will henceforth be called the Rayleigh–Ritz–Galerkin (RRG) eigenvalue and eigenfunction.

Using the equality of the two characterizations, we immediately have the following basic result.

THEOREM 8.2

If S is any n-dimensional subspace of $PC_0^{1,2}(I)$ and $\tilde{\lambda}_j$, $1 \leq j \leq n$, are the corresponding approximate eigenvalues of (8.1)–(8.2), then

$$\lambda_j \leq \tilde{\lambda}_j, \qquad 1 \leq j \leq n. \tag{8.14}$$

This result yields an upper bound for the eigenvalues. In Section 8.3, we will also give a lower bound and hence an a priori error estimate.

Using the basis functions for the subspaces $L_0(\Delta)$, $H_0(\Delta)$, and $S_0(\Delta)$ described in Chapter 7, we obtain sparse matrices which for one-dimensional problems are actually band matrices. Moreover, as shown in Chapter 7, these matrices are well-conditioned.

We now give the numerical results of [8.1] for the very simple problem

$$-D^2u = \pi^2\lambda u, \qquad 0 < x < 1, \tag{8.15}$$

$$u(0) = u(1) = 0. \tag{8.16}$$

For $H_0(\Delta)$, a search was made for the zeroes of det $(A - \mu B)$ using regula falsi and a special code for seven-diagonal determinants. For $S_0(\Delta)$, the eigenvalues were computed by applying a standard eigenvalue subroutine to the not necessarily symmetric matrix $B^{-1}A$. Moreover, all the partitions were chosen to be uniform.

h	$\tilde{\lambda}_1(H_0) - \lambda_1$	$\tilde{\lambda}_2(H_0) - \lambda_2$	$\tilde{\lambda}_3(H_0) - \lambda_3$	$\tilde{\lambda}_1(S_0) - \lambda_1$	$\tilde{\lambda}_2(S_0) - \lambda_2$	$\tilde{\lambda}_3(S_0) - \lambda_3$
$\frac{1}{3}$	$.32 \times 10^{-6}$	$.43 \times 10^{-4}$	$.12 \times 10^{-2}$	$.5 \times 10^{-2}$		
$\frac{1}{4}$	$.64 \times 10^{-7}$	$.11 \times 10^{-4}$	$.17 \times 10^{-3}$		$.52 \times 10^{-4}$	$.27 \times 10^{-2}$
$\frac{1}{5}$	$.18 \times 10^{-7}$	$.34 \times 10^{-5}$	$.58 \times 10^{-4}$		$.11 \times 10^{-4}$	$.48 \times 10^{-3}$
$\frac{1}{6}$	$.64 \times 10^{-8}$	$.13 \times 10^{-5}$	$.24 \times 10^{-4}$		$.28 \times 10^{-5}$	$.12 \times 10^{-3}$
$\frac{1}{7}$	$.27 \times 10^{-8}$	$.54 \times 10^{-6}$	$.11 \times 10^{-4}$		$.10 \times 10^{-5}$	$.38 \times 10^{-4}$
$\frac{1}{8}$	$.12 \times 10^{-8}$	$.26 \times 10^{-6}$	$.42 \times 10^{-5}$		$.40 \times 10^{-6}$	$.15 \times 10^{-4}$
$\frac{1}{9}$	$.68 \times 10^{-9}$	$.13 \times 10^{-6}$	$.28 \times 10^{-5}$		$.20 \times 10^{-6}$	$.64 \times 10^{-5}$

For the case of $L_0(\Delta)$, where Δ is a uniform partition of mesh length h, we obtain

$$A \equiv h^{-1}\begin{bmatrix} 2 & -1 & & 0 \\ -1 & \ddots & \ddots & \\ & \ddots & \ddots & -1 \\ 0 & & -1 & 2 \end{bmatrix} \quad \text{and} \quad B \equiv \frac{h}{6}\begin{bmatrix} 4 & 1 & & 0 \\ 1 & \ddots & \ddots & \\ & \ddots & \ddots & 1 \\ 0 & & 1 & 4 \end{bmatrix},$$

and hence $B = h(I - (h/6)A)$. Therefore, if $A\boldsymbol{\beta} = \tilde{\lambda}B\boldsymbol{\beta}$, we have

$$hA\boldsymbol{\beta} = \tilde{\lambda}h^2\left(I - \frac{h}{6}A\right)\boldsymbol{\beta}$$

or

$$hA\boldsymbol{\beta} = \tilde{\lambda}\left(1 + \tilde{\lambda}\frac{h^2}{6}\right)^{-1}h^2\boldsymbol{\beta} \equiv \mu\boldsymbol{\beta}.$$

But it is easily verified, as in Section 7.2, that the eigenvalues of hA are

$$\mu_j = 2h^{-2}(1 - \cos j\pi h), \qquad 1 \leq j \leq h^{-1},$$

and therefore the eigenvalues $\tilde{\lambda}_j$ are given by

$$\tilde{\lambda}_j = 6h^2(1 - \cos j\pi h)(2 + \cos j\pi h)^{-1}.$$

Moreover, it is easily verified that

$$\tilde{\lambda}_1\left(1 + \frac{h^2}{6}\tilde{\lambda}_1\right)^{-1} < \lambda_1 = \pi^2 < \tilde{\lambda}_1.$$

These numerical results show that for this special case the Rayleigh–Ritz–Galerkin procedure yields approximate eigenvalues over $L_0(\Delta)$ which are second-order accurate, and approximate eigenvalues over $H_0(\Delta)$ and $S_0(\Delta)$ which are sixth-order accurate. In Section 8.3, we will prove that this is true in general for sufficiently smooth eigenfunctions.

8.3 ERROR ANALYSIS

In this section, we prove a priori error bounds for the Rayleigh–Ritz–Galerkin approximations to the first eigenvalue and eigenfunction. Our method of proof follows [8.1] and [8.3]. See Exercise (8.2) for corresponding results about the approximations to the higher eigenvalues and eigenfunctions.

We begin by discussing the approximate eigenvalue.

THEOREM 8.3

Let S be any finite-dimensional subspace of $PC_0^{1,2}(I)$, $\{\tilde{\lambda}_1, \tilde{u}_1\}$ the RRG eigenvalue and eigenfunction in S corresponding to $\{\lambda_1, u_1\}$ and ϑ any linear mapping of $PC_0^{1,2}(I)$ into S. If $b[u_1 - \vartheta u_1] < 1$, then

$$\lambda_1 \leq \tilde{\lambda}_1 \leq R[\vartheta u_1] \leq \lambda_1 + a[u_1 - \vartheta u_1](1 - b^{1/2}[u_1 - \vartheta u_1])^{-2}. \tag{8.17}$$

Proof. Let $e_1 \equiv \vartheta u_1 - u_1$ and $P_r e_1 \equiv \delta_1$, where P_r denotes the orthogonal projection mapping of $PC_0^{1,2}(I)$ onto the one-dimensional subspace spanned by u_1 with respect to the inner product $b(w, v)$.

We now make the important observation that if $u \in \text{span}(u_1)$, i.e., u is in the one-dimensional subspace spanned by u_1, and $b(v, u_1) = 0$, then

$$b[u + v] = b[u] + b[v] \quad \text{and} \quad a[u + v] = a[u] + a[v]. \tag{8.18}$$

In fact,

$$\begin{aligned} b[u + v] &= b(u + v, u + v) = b[u] + b[v] + 2b[u, v] \\ &= b[u] + b[v] \end{aligned}$$

and

$$\begin{aligned} a[u + v] &= a[u] + a[v] + 2a(u, v) = a[u] + a[v] + \lambda_1 b(u, v) \\ &= a[u] + a[v]. \end{aligned}$$

Hence, since $\vartheta u_1 = (u_1 + \delta_1) + (e_1 - \delta_1)$, we have

$$\begin{aligned} b[\vartheta u_1] &= b[u_1 + \delta_1] + b[e_1 - \delta_1], \\ a[\vartheta u_1] &= a[u_1 + \delta_1] + a[e_1 - \delta_1], \end{aligned}$$

and

$$a[e_1] = a[\delta_1] + a[e_1 - \delta_1].$$

Thus,

$$\begin{aligned} R[\vartheta u_1] &= a[\vartheta u_1](b[\vartheta u_1])^{-1} = (a[u_1 + \delta_1] + a[e_1 - \delta_1])(b[\vartheta u_1])^{-1} \\ &\leq (a[u_1 + \delta_1])(b[u_1 + \delta_1])^{-1} + (a[e_1 - \delta_1])(b[\vartheta u_1])^{-1} \\ &= \lambda_1 + (a[e_1 - \delta_1])(b[\vartheta u_1])^{-1}, \end{aligned}$$

where we have used the fact that

$$b[u_1 + \delta_1] = b[\vartheta u_1] - b[e_1 - \delta_1] \leq b[\vartheta u_1].$$

Hence,

$$R[\vartheta u_1] \leq \lambda_1 + (a[e_1])(b[\vartheta u_1])^{-1}. \tag{8.19}$$

But by the triangle inequality,

$$b^{1/2}[u_1] \leq b^{1/2}[\vartheta u_1] + b^{1/2}[e_1]$$

or

$$b[\vartheta u_1] \geq (b^{1/2}[u_1] - b^{1/2}[e_1])^2 = (1 - b^{1/2}[e_1])^2. \tag{8.20}$$

Using (8.20) to bound the right-hand side of (8.19), we obtain (8.17). Q.E.D.

We now turn to a derivation of an a priori errror bound for the first eigenfunction.

THEOREM 8.4

Let $0 < \lambda_1 < \lambda_2 \leq \ldots$, S be any finite-dimensional subspace of $PC_0^{1,2}(I)$, and $\{\tilde{\lambda}_1, \tilde{u}_1\}$ be the RRG eigenvalue and eigenfunction in S corresponding to $\{\lambda_1, u_1\}$. If $\tilde{\lambda}_1 < \lambda_2$, then

$$a[u_1 - \tilde{u}_1] \leq \tilde{\lambda}_1 - \lambda_1 + 2\lambda_1[1 - \{1 - (\tilde{\lambda}_1 - \lambda_1)(\lambda_2 - \lambda_1)^{-1}\}^{1/2}]. \tag{8.21}$$

Proof. We normalize the first RRG eigenfunction $\tilde{u}_1$ by $b[\tilde{u}_1] = 1$, $(a[\tilde{u}_1] = \tilde{\lambda}_1)$, and $b^2(u_1, \tilde{u}_1) \equiv 1 - \sigma_1^2 \geq 0$. Thus,

$$\begin{aligned} b[u_1 - \tilde{u}_1] &= b[u_1] - 2b(u_1, \tilde{u}_1) + b[\tilde{u}_1] = 2(1 - b(u_1, \tilde{u}_1)) \\ &= 2(1 - (1 - \sigma_1^2)^{1/2}). \end{aligned}$$

Moreover, since $\tilde{u}_1 = b(u_1, \tilde{u}_1)u_1 + \eta$, where $b(\eta, u_1) = 0$, we have

$$\sigma_1^2 = 1 - b^2(u_1, \tilde{u}_1) = b(\eta, \tilde{u}_1) = b(\eta, \eta) = b[\eta]. \tag{8.22}$$

Let $a_1[w] \equiv a[w] - \lambda_1 b[w]$ and $a_2[w] \equiv a[w] - \lambda_2 b[w]$ for all $w \in PC_0^{1,2}(I)$. Since $a_2[w] + (\lambda_2 - \lambda_1)b[w] = a_1[w]$ for all $w \in PC_0^{1,2}(I)$, and $a_2[w] \geq 0$ for all w such that $b(w, u_1) = 0$, we have

$$(\lambda_2 - \lambda_1)b[\eta] \leq a_1[\eta] = a_1[\tilde{u}_1] = \tilde{\lambda}_1 - \lambda_1, \tag{8.23}$$

where we have used (8.18) with $u = b(u_1, \tilde{u}_1)u_1$ and $v = \eta$. From (8.23), we have

$$b[\eta] \leq (\tilde{\lambda}_1 - \lambda_1)(\lambda_2 - \lambda_1)^{-1}, \tag{8.24}$$

and from (8.22) and (8.24) we have

$$\sigma_1^2 \leq (\tilde{\lambda}_1 - \lambda_1)(\lambda_2 - \lambda_1)^{-1}. \tag{8.25}$$

Moreover,

$$\begin{aligned} \tilde{\lambda}_1 - \lambda_1 &= a_1[\tilde{u}_1] - a_1[u_1] = a_1[u_1 + (\tilde{u}_1 - u_1)] - a_1[u_1] \\ &= 2(a(u_1, \tilde{u}_1 - u_1) - \lambda_1 b(u_1, \tilde{u}_1 - u_1)) + a_1[\tilde{u}_1 - u_1] \\ &= a_1[\tilde{u}_1 - u_1] = a[\tilde{u}_1 - u_1] - \lambda_1 b[\tilde{u}_1 - u_1]. \end{aligned}$$

Hence

$$\begin{aligned} a[\tilde{u}_1 - u_1] &= (\tilde{\lambda}_1 - \lambda_1) + \lambda_1 b[\tilde{u}_1 - u_1] \\ &= (\tilde{\lambda}_1 - \lambda_1) + 2\lambda_1(1 - (1 - \sigma_1^2)^{1/2}), \end{aligned} \tag{8.26}$$

and we obtain (8.21) by using the inequality (8.25) to bound the right-hand side of (8.26). Q.E.D.

COROLLARY

Let $0 < \lambda_1 < \lambda_2 \leq \ldots$, let S be a finite-dimensional subspace of $PC_0^{1,2}(I)$, and let $\{\tilde{\lambda}_1, \tilde{u}_1\}$ be the RRG eigenvalue and eigenfunction in S corresponding to $\{\lambda_1, u_1\}$. If $\tilde{\lambda}_1 < \lambda_2$ and ϑ is any linear mapping of $PC_0^{1,2}(I)$ into S such that $\xi_1 \equiv b[u_1 - \vartheta u_1] < 1$, then

$$a[u_1 - \tilde{u}_1] \leq \eta_1(1 - \xi_1^{1/2})^{-2} + 2\lambda_1[1 - \{1 - \eta_1(1 - \xi_1^{1/2})^{-2}(\lambda_2 - \lambda_1)^{-1}\}^{1/2}], \tag{8.27}$$

where $\eta_1 \equiv a[u_1 - \vartheta u_1]$.

Combining the results of Theorem 8.3 and the Corollary to Theorem 8.4 with the approximation theory results of Chapters 2, 3, and 6, we obtain the following a priori bounds. Using the results of Theorem 2.5, we may prove the following theorem.

THEOREM 8.5

Let $u_1 \in PC^{2,2}(I) \cap PC_0^{1,2}(I)$ and $\{\tilde{\lambda}_1, \tilde{u}_1\}$ be the RRG eigenvalue and eigenfunction in $L_0(\Delta)$ corresponding to $\{\lambda_1, u_1\}$. If $\xi_1 \equiv \pi^{-4}h^4 \|D^2u_1\|_2 \|r\|_\infty < 1$, then

$$\lambda_1 \leq \tilde{\lambda}_1 \leq \lambda_1 + \eta_1(1 - \xi_1^{1/2})^{-2}, \tag{8.28}$$

where

$$\eta_1 \equiv (\|p\|_\infty + \pi^{-2}h^2 \|q\|_\infty)\pi^{-2}h^2 \|D^2u_1\|_2^2.$$

Moreover, if in addition $\lambda_1 < \lambda_2$ and Δ is such that $\tilde{\lambda}_1 < \lambda_2$, then

$$a[u_1 - \tilde{u}_1] \leq \eta_1(1 - \xi_1^{1/2})^{-2} + 2\lambda_1[1 - \{1 - \eta_1(1 - \xi_1^{1/2})^{-2}(\lambda_2 - \lambda_1)^{-1}\}^{1/2}]. \tag{8.29}$$

It follows that if $u_1 \in PC^{2,2}(I) \cap PC_0^{1,2}(I)$ and $S \equiv L_0(\Delta)$, then $\tilde{u}_1$ is a first-order accurate approximation to u_1 with respect to the norm $a^{1/2}[\cdot]$ and $\tilde{\lambda}_1$ is a second-order accurate approximation to λ_1. Using the results of Theorem 3.5, we may prove the following theorem.

THEOREM 8.6

Let $u_1 \in PC^{4,2}(I) \cap PC_0^{1,2}(I)$ and $\{\tilde{\lambda}_1, \tilde{u}_1\}$ be the RRG eigenvalue and eigenfunction in $H_0(\Delta)$ corresponding to $\{\lambda_1, u_1\}$. If $\xi_1 \equiv \pi^{-8}h^8 \|D^4u_1\|_2^2 \|r\|_\infty < 1$, then

$$\lambda_1 \leq \tilde{\lambda}_1 \leq \lambda_1 + \eta_1(1 - \xi_1^{1/2})^{-2}, \tag{8.30}$$

where

$$\eta_1 \equiv (\|p\|_\infty + \pi^{-2}h^2\|q\|_\infty)\pi^{-6}h^6\|D^4u_1\|_2^2.$$

Moreover, if in addition $\lambda_1 < \lambda_2$ and Δ is such that $\tilde{\lambda}_1 < \lambda_2$, then

$$a[u_1 - \tilde{u}_1] \leq \eta_1(1 - \xi_1^{1/2})^{-2} + 2\lambda_1[1 - \{1 - \eta_1(1 - \xi_1^{1/2})^{-2}(\lambda_2 - \lambda_1)^{-1}\}^{1/2}]. \tag{8.31}$$

Finally, using the results of Theorem 6.7, we may prove the following theorem.

THEOREM 8.7

Let $u_1 \in PC^{4,2}(I) \cap PC_0^{1,2}(I)$ and $\{\tilde{\lambda}_1, \tilde{u}_1\}$ be the RRG eigenvalue and eigenfunction in $S_0(\Delta)$ corresponding to $\{\lambda_1, u_1\}$. If $\xi_1 \equiv 4\pi^{-8}h^8\|D^4u_1\|_2\|r\|_\infty < 1$, then

$$\lambda_1 \leq \tilde{\lambda}_1 \leq \lambda_1 + \eta_1(1 - \xi_1^{1/2})^{-2}, \tag{8.32}$$

where

$$\eta_1 \equiv 4(\|p\|_\infty + \pi^{-2}h^2\|q\|_\infty)\pi^{-6}h^6\|D^4u_1\|_2^2.$$

Moreover, if in addition $\lambda_1 < \lambda_2$ and Δ is such that $\tilde{\lambda}_1 < \lambda_2$, then

$$a[u_1 - \tilde{u}_1] \leq \eta_1(1 - \xi_1^{1/2})^{-2} + 2\lambda_1[1 - \{1 - \eta_1(1 - \xi_1^{1/2})^{-2}(\lambda_2 - \lambda_1)^{-1}\}^{1/2}]. \tag{8.33}$$

We remark that if $u \in PC^{4,2}(I) \cap PC_0^{1,2}(I)$ and $S \equiv H_0(\Delta)$ or $S \equiv S_0(\Delta)$, then $\tilde{u}_1$ is a third-order accurate approximation to u_1 with respect to the norm $a^{1/2}[\cdot]$ and $\tilde{\lambda}_1$ is a sixth-order accurate approximation to λ_1.

EXERCISES FOR CHAPTER 8

(8.1) Show that if $a(u, v)$ is strongly coercive, $\{\tilde{\lambda}_1, \tilde{u}_1\}$ are the RRG eigenvalue and eigenfunction in $L_0(\Delta)$ corresponding to $\{\lambda_1, u_1\}$, and

$$2\|r\|_\infty^{1/2}\mu^{3/2}\pi^{-2}\gamma^{-1/2}\Gamma h^2\|D^2u_1\|_2 < 1,$$

then

$$\tilde{\lambda}_1(1 - 2\|r\|_\infty^{1/2}\mu^{3/2}\pi^{-2}\gamma^{-1/2}\Gamma h^2\|D^2u_1\|_2) \leq \lambda_1 \leq \tilde{\lambda}_1$$

(cf. [8.12].)

(8.2) Prove the following extension of Theorems 8.3 and 8.4. Let $u_1, \ldots, u_k$ denote the first k eigenfunctions normalized so that $b(u_i, u_j) = \delta_{ij}$ and S be a finite-dimensional subspace of $PC_0^{1,2}(I)$ of dimension greater than k.

If there exist k functions $w_1, \ldots, w_k \in S$ such that

$$\sum_{i=1}^{k} b[w_i - u_i] < 1,$$

then

$$\lambda_j \leq \tilde{\lambda}_j \leq \left(\sum_{i=1}^{j} a[w_i - u_i]\right)\left[1 - \left(\sum_{i=1}^{j} b[w_i - u_i]\right)^{1/2}\right]^2,$$

for all $1 \leq j \leq k$. Moreover, if in addition $0 < \lambda_1 < \lambda_2 < \ldots < \lambda_k$, there exists a positive constant, K, such that

$$a[u_j - \tilde{u}_j] \leq K \sum_{i=1}^{j} (\tilde{\lambda}_i - \lambda_i), \qquad 1 \leq j \leq k.$$

(Hint: cf. [8.1] and [8.3]). (See [8.10] and [8.11] for sharp error bounds for the RRG eigenfunctions with respect to the L^2-norm.)

(8.3) Prove analogues of the results of Section 8.2 for the spaces

$$S_0(2m-1, \Delta, z) \equiv \{\phi \in S(2m-1, \Delta, z) \mid \phi(0) = \phi(1) = 0\}$$

defined in Exercise (4.12).

REFERENCES FOR CHAPTER 8

[8.1] BIRKHOFF, G., C. DEBOOR, B. SWARTZ, and B. WENDROFF, Rayleigh–Ritz approximation by piecewise cubic polynomials. *SIAM J. Numer. Anal.* **3**, 188–203 (1966).

[8.2] BRAUER, F., Singular self-adjoint boundary value problems for the differential equation $Lx = \lambda Mx$. *Trans. Amer. Math. Soc.* **88**, 331–345 (1958).

[8.3] CIARLET, P. G., M. H. SCHULTZ, and R. S. VARGA, Numerical methods of high-order accuracy for nonlinear boundary value problems. III. Eigenvalue problems. *Numer. Math.* **12**, 120–133 (1968).

[8.4] COLLATZ, L. *The Numerical Treatment of Differential Equations*, 3rd ed. Springer-Verlag, Berlin (1960).

[8.5] COURANT, R., and D. HILBERT, *Methods of Mathematical Physics.* Interscience, New York (1962).

[8.6] GOULD, S. H., *Variational Methods for Eigenvalue Problems.* University of Toronto Press, Toronto (1966).

[8.7] KAMKE, E. A., Über die definiten selbstadjungierten Eigenwertaufgaben bei gewöhnlichen linearen Differentialgleichungen. II, III. *Math. Z.* **46**, 231–286 (1940).

[8.8] KAMKE, E. A., Über die definiten selbstadjungierten Eigenwertaufgaben bei gewöhnlichen linearen Differentialgleichungen. IV. *Math. Z.* **48**, 67–100 (1942).

[8.9] KANTOROVICH, L. V., and V. I. KRYLOV, *Approximate Methods of Higher Analysis*. Interscience, New York (1958).

[8.10] PIERCE, J. G., and R. S. VARGA, Higher order convergence results for the Rayleigh–Ritz method applied to eigenvalue problems. I. Estimates relating Rayleigh–Ritz and Galerkin approximations to eigenfunctions. *SIAM J. Numer. Anal.* **9**, 137–151 (1972).

[8.11] SCHULTZ, M. H., L^2-error bounds for the Rayleigh–Ritz–Galerkin method. *SIAM J. Numer. Anal.* **8**, 737–748 (1971).

[8.12] WENDROFF, B., Bounds for eigenvalues of some differential operators by the Rayleigh-Ritz method. *Math. Comp.* **19**, 218–224 (1965).

9 SEMI-DISCRETE GALERKIN PROCEDURES FOR PARABOLIC EQUATIONS

9.1 LINEAR PROBLEMS

In this chapter, we discuss the use of the Galerkin procedure to "discretize" the space variables in initial boundary value problems for linear and semilinear parabolic differential equations. In this section, we consider the linear problem

$$(9.1) \qquad D_t u = D_x[p(x)D_x u] - q(x)u + f(x), \qquad 0 < x < 1, \quad t > 0,$$

subject to the initial condition

$$(9.2) \qquad u(x, 0) = u_0(x), \qquad 0 < x < 1,$$

and the boundary conditions

$$(9.3) \qquad u(0, t) = u(1, t) = 0, \qquad t \geq 0,$$

where $p(x)$ and $q(x) \in PC^{0,\infty}(I)$, $f(x) \in PC^{0,2}(I)$, $u_0(x) \in PC_0^{1,2}(I)$, and such that there exist two positive constants, γ and μ, for which

$$(9.4) \qquad \begin{aligned} \gamma \| D_x u \|_2^2 &\leq \int_0^1 \{p(x)[D_x u]^2 + q(x)[u]^2\}\, dx = a(u, u) \\ &\leq \mu \| D_x u \|_2^2, \end{aligned}$$

for all $u \in PC_0^{1,2}(I)$. See [9.2], [9.3], [9.6], [9.7], and [9.8] for discussions of more general problems.

We say that u is a *generalized solution* of (9.1)–(9.3) if and only if $u(x, t) \in$

$PC_0^{1,2}(I)$ for all $t \geq 0$, $u(x, 0) = u_0(x)$, and

$$(9.5) \qquad (D_t u, v)_2 + a(u, v) = (f, v)_2, \qquad t > 0, \text{ for all } v \in PC_0^{1,2}(I).$$

Integrating by parts and using the Gronwall Inequality (cf. [9.1]), we can prove the following standard result; cf. [9.5].

THEOREM 9.1

If u is a classical solution of (9.1)–(9.3), then it is a generalized solution. Moreover, (9.1)–(9.3) has at most one generalized solution.

Throughout this section, we will assume the generalized solution exists.

To define a "semi-discrete Galerkin" approximation to the generalized solution, u, of (9.1)–(9.3), we let S be a finite-dimensional subspace of $PC_0^{1,2}(I)$ spanned by the basis functions $\{B_i(x)\}_{i=1}^n$ and seek an approximation $u_S(x, t)$ of the form

$$u_S(x, t) = \sum_{i=1}^{n} \beta_i(t) B_i(x).$$

The coefficients $\{\beta_i(t)\}_{i=1}^n$ are functions of time which are determined as the solution of the linear system of ordinary differential equations

$$(9.6) \qquad (D_t u_S, B_i)_2 + a(u_S, B_i) = (f, B_i), \qquad 1 \leq i \leq n, \text{ for all } t > 0,$$

and

$$(9.7) \qquad (u_S(0), B_i)_2 = (u_0, B_i)_2, \qquad 1 \leq i \leq n.$$

Expressing u_S in terms of the basis functions, we obtain the equivalent system

$$(9.8) \qquad BD_t\boldsymbol{\beta}(t) + A\boldsymbol{\beta}(t) = \mathbf{k}, \qquad t > 0,$$

$$(9.9) \qquad B\boldsymbol{\beta}(0) = \mathbf{g},$$

where $B \equiv [b_{ij}] \equiv [(B_i, B_j)_2]$, $A \equiv [a_{ij}] \equiv [a(B_i, B_j)]$, $\mathbf{k} \equiv [k_i] \equiv [(f, B_i)_2]$, and $\mathbf{g} \equiv [g_i] \equiv [(u_0, B_i)_2]$.

Since B is symmetric, positive definite and hence nonsingular, it follows from a standard result in ordinary differential equations (cf. [9.1]), that (9.8)–(9.9) has a unique solution $\boldsymbol{\beta}^*(t)$. Moreover, the solution $\boldsymbol{\beta}^*(t)$ can be expressed analytically as

$$(9.10) \qquad \boldsymbol{\beta}^*(t) = e^{-tB^{-1}A}B^{-1}\mathbf{g} + \int_0^t e^{(s-t)B^{-1}A}B^{-1}\mathbf{k}\, ds, \qquad t \geq 0.$$

THEOREM 9.2

Under the preceding hypotheses, the semi-discrete Galerkin procedure is well-defined for linear problems.

We now turn to the question of a priori error bounds for this procedure.

THEOREM 9.3

If $\tilde{u}_S(x, t)$ is the orthogonal projection of $u(x, t)$ onto S with respect to the inner product $a(u, v)$ for all $t \geq 0$, then

$$\text{(9.11)} \quad \begin{aligned} \|u(t) - u_S(t)\|_2 \leq \|u(t) - \tilde{u}_S(t)\|_2 + [\|u_S(0) - \tilde{u}_S(0)\|_2^2 \\ + t(2\gamma\pi^2)^{-1} \sup_{0 \leq s \leq t} \|D_t(u - \tilde{u}_S)(s)\|_2^2]^{1/2}, \qquad t \geq 0. \end{aligned}$$

Proof. From Theorem 7.13 we have that for each $t \geq 0$,

$$\text{(9.12)} \qquad a(u(t), B_i) = a(\tilde{u}_S(t), B_i), \qquad 1 \leq i \leq n.$$

It follows from equation (9.6) that for all $w \in S$, we have

$$\text{(9.13)} \quad \begin{aligned} (D_t(u - \tilde{u}_S)(t), w)_2 &= (f, w)_2 - a(u(t), w) - (D_t\tilde{u}_S(t), w)_2 \\ &= (D_t u_S(t), w)_2 + a(u_S(t), w) - a(\tilde{u}_S(t), w) \\ &\quad - (D_t\tilde{u}_S(t), w)_2, \qquad t > 0. \end{aligned}$$

Thus,

$$\text{(9.14)} \quad \begin{aligned} &(D_t(u - \tilde{u}_S)(t), w)_2 \\ &\quad = (D_t(u_S - \tilde{u}_S)(t), w)_2 + a((u_S - \tilde{u}_S)(t), w), \qquad t > 0. \end{aligned}$$

Choosing $w \equiv u_S(t) - \tilde{u}_S(t)$ in (9.14), we obtain

$$\text{(9.15)} \quad \begin{aligned} &(D_t(u - \tilde{u}_S)(t), (u_S - \tilde{u}_S)(t))_2 \\ &\quad = (D_t(u_S - \tilde{u}_S)(t), (u_S - \tilde{u}_S)(t))_2 + a((u_S - \tilde{u}_S)(t), (u_S - \tilde{u}_S)(t)) \\ &\quad = \tfrac{1}{2} D_t \|(u_S - \tilde{u}_S)(t)\|_2^2 + a((u_S - \tilde{u}_S)(t), (u_S - \tilde{u}_S)(t)). \end{aligned}$$

But

$$\begin{aligned} &(D_t(u - \tilde{u}_S)(t), (u_S - \tilde{u}_S)(t))_2 \\ &\qquad \leq (4\gamma\pi^2)^{-1} \|D_t(u - \tilde{u}_S)(t)\|_2^2 + \gamma\pi^2 \|(u_S - \tilde{u}_S)(t)\|_2^2, \end{aligned}$$

and hence

$$\text{(9.16)} \quad \begin{aligned} &\tfrac{1}{2} D_t \|(u_S - \tilde{u}_S)(t)\|_2^2 \\ &\quad \leq (4\gamma\pi^2)^{-1} \|D_t(u - \tilde{u}_S)(t)\|_2^2 + \gamma\pi^2 \|(u_S - \tilde{u}_S)(t)\|_2^2 \\ &\qquad - a((u_S - \tilde{u}_S)(t), (u_S - \tilde{u}_S)(t)) \\ &\quad \leq (4\gamma\pi^2)^{-1} \|D_t(u - \tilde{u}_S)(t)\|_2^2. \end{aligned}$$

Integrating both sides of the inequality (9.16) from 0 to t, we obtain

$$\text{(9.17)} \quad \|(u_S - \tilde{u}_S)(t)\|_2^2 \leq \|(u_S - \tilde{u}_S)(0)\|_2^2 + t(2\gamma\pi^2)^{-1} \sup_{0 \leq s \leq t} \|D_t(u - \tilde{u}_S)(s)\|_2^2,$$

and the result follows from the triangle inequality. Q.E.D.

Since $D_t\tilde{u}_S$ is the orthogonal projection of $D_t u$ onto S with respect to the inner product $a(u, v)$, i.e.,

$$a(D_t\tilde{u}_S, w) = a(D_t u, w), \qquad \text{for all } w \in S,$$

we may use the results of Chapters 6 and 7 to bound the right-hand side of (9.11).

THEOREM 9.4

If $a(u, v)$ is strongly coercive and $u \in PC^{3,\infty}(I \times (0, \infty))$, then

$$\begin{aligned} \|(u - u_L)(t)\|_2 &\leq \gamma^{-1/2}\mu^{3/2}\pi^{-2}\Gamma h^2 \|D_x^2 u(t)\|_2 \\ &\quad + [(1 + \gamma^{-1/2}\mu^{3/2}\Gamma)^2 \|D_x^2 u_0\|_2^2 + t(2\gamma\pi^2)^{-1}\gamma^{-1}\mu^3\Gamma^2 \\ &\quad \times \sup_{0 \leq s \leq t} \|D_t D_x^2 u(s)\|_2^2]^{1/2}\pi^{-2}h^2, \qquad t \geq 0, \end{aligned} \tag{9.18}$$

where u_L is the semi-discrete Galerkin approximation to u over $L_0(\Delta)$.

Proof. By the results of Theorems 6.5 and 7.15,

$$\|(u - \tilde{u}_L)(t)\|_2 \leq \gamma^{-1/2}\mu^{3/2}\Gamma\pi^{-2}h^2 \|D_x^2 u(t)\|_2 \tag{9.19}$$

and hence

$$\begin{aligned} \|u_L(0) - \tilde{u}_L(0)\|_2 &\leq \|u_0 - u_L(0)\|_2 + \|u_0 - \tilde{u}_L(0)\|_2 \\ &\leq (1 + \gamma^{-1/2}\mu^{3/2}\Gamma)\pi^{-2}h^2 \|D_x^2 u_0\|_2. \end{aligned} \tag{9.20}$$

Moreover,

$$\begin{aligned} &t(2\gamma\pi^2)^{-1} \sup_{0 \leq s \leq t} \|D_t(u - \tilde{u}_L)(s)\|_2^2 \\ &\qquad \leq t(2\gamma\pi^2)^{-1}\gamma^{-1}\mu^3\pi^{-4}\Gamma^2 h^4 \sup_{0 \leq s \leq t} \|D_t D_x^2 u(s)\|_2^2. \end{aligned} \tag{9.21}$$

Inequality (9.18) now follows by using (9.19)–(9.21) to bound the right-hand side of (9.11) Q.E.D.

Likewise, we may prove the following result.

THEOREM 9.5

If $a(u, v)$ is strongly coercive and $u \in PC^{5,\infty}(I \times (0, \infty))$, then

$$\begin{aligned} \|(u - u_H)(t)\|_2 &\leq \gamma^{-1/2}\mu^{3/2}\pi^{-4}\Gamma h^4 \|D_x^4 u(t)\|_2 \\ &\quad + [(1 + \gamma^{-1/2}\mu^{3/2}\Gamma)^2 \|D_x^4 u_0\|_2^2 + t(2\gamma\pi^2)^{-1}\gamma^{-1}\mu^3\Gamma^2 \\ &\quad \times \sup_{0 \leq s \leq t} \|D_t D_x^4 u(s)\|_2^2]^{1/2}\pi^{-4}h^4 \\ &\equiv \Omega_3(t)\pi^{-4}h^4, \qquad t > 0 \end{aligned} \tag{9.22}$$

and

$$\| (u - u_S)(t) \|_2 \leq 4\Omega_3(t)\pi^{-4}h^4, \qquad t > 0, \tag{9.23}$$

where u_H and u_S denote the semi-discrete Galerkin approximations to u over $H_0(\Delta)$ and $S_0(\Delta)$, respectively.

Thus, we have shown that under suitable hypotheses, the semi-discrete Galerkin approximation to u over $L_0(\Delta)$ is second-order accurate with respect to the L^2-norm, while the semi-discrete Galerkin approximations to u over $H_0(\Delta)$ and $S_0(\Delta)$ are both fourth-order accurate with respect to the L^2-norm.

9.2 SEMILINEAR PROBLEMS

In this section, we consider the semilinear problem

$$D_t u = D_x[p(x)D_x u] - q(x)u + f(x, u), \qquad 0 < x < 1, \quad t > 0, \tag{9.24}$$

subject to the initial condition

$$u(x, 0) = u_0(x), \qquad 0 < x < 1, \tag{9.25}$$

and the boundary conditions

$$u(0, t) = u(1, t) = 0, \qquad t \geq 0, \tag{9.26}$$

where the coefficients $p(x)$, $q(x)$, and $u_0(x)$ satisfy the hypotheses of Section 9.1, $f(x, u)$ and $(\partial f/\partial u)(x, u)$ are continuous on $I \times (-\infty, \infty)$, $|(\partial f/\partial u)(x, u)| \leq B$ for all $(x, u) \in I \times (-\infty, \infty)$, and

$$\frac{\partial f}{\partial u}(x, u) \leq \lambda < \Lambda \equiv \inf_{w \in PC_0^{1,2}(I)} \frac{a(w, w)}{(w, w)_2}.$$

We say that u is a *generalized solution* of (9.24)–(9.26) if and only if $u(x, t) \in PC_0^{1,2}(I)$ for all $t \geq 0$, $u(x, 0) = u_0(x)$, and

$$(D_t u, v)_2 + a(u, v) = (f(u), v)_2, \qquad t > 0, \text{ for all } v \in PC_0^{1,2}(I). \tag{9.27}$$

As in the linear case, we may prove the following standard result; cf. [9.5].

THEOREM 9.6

If u is a classical solution of (9.24)–(9.26), then it is a generalized solution. Moreover, (9.24)–(9.26) has at most one generalized solution.

Throughout this section, we will assume that the generalized solution exists. Given a finite-dimensional subspace S of $PC_0^{1,2}(I)$, we find the semi-discrete Galerkin approximation

$$u_S(x, t) \equiv \sum_{i=1}^{n} \beta_i(t) B_i(x)$$

by solving the nonlinear system of ordinary differential equations,

$$(D_t u_S, w)_2 + a(u_S, w) = (f(u_S), w)_2, \qquad t > 0, \text{ for all } w \in S, \tag{9.28}$$

$$(u_S(0), w)_2 = (u_0, w)_2, \qquad \text{for all } w \in S, \tag{9.29}$$

or equivalently

$$BD_t\boldsymbol{\beta}(t) + A\boldsymbol{\beta}(t) = \mathbf{f}(\boldsymbol{\beta}), \ t > 0, \tag{9.30}$$

$$B\boldsymbol{\beta}(0) = \mathbf{g}, \tag{9.31}$$

where B and A are the previously defined matrices and

$$\mathbf{f}(\boldsymbol{\beta}) \equiv [f_i(\boldsymbol{\beta})] \equiv \left[\left(f\left(\sum_{i=1}^{n} \beta_i B_i\right), B_i\right)\right].$$

Since B is nonsingular, it follows from a standard result in ordinary differential equations (cf. [9.11]), that (9.30)–(9.31) has a unique solution.

THEOREM 9.7

Under the preceding hypotheses, the semi-discrete Galerkin method is well-defined for semilinear problems.

We turn now to the question of a priori error bounds for this procedure.

THEOREM 9.8

If, for all $t \geq 0$, $\tilde{u}_S(x, t)$ is the RRG approximation in S of the solution, $v(x)$, of the nonlinear two-point boundary problem

$$\begin{aligned} &-D_x[p(x)D_x v] + q(x)v - f(x, v) \\ &\quad = -D_x[p(x)D_x u(x, t)] + q(x)u(x, t) - f(x, u(x, t)), \qquad 0 < x < 1, \end{aligned} \tag{9.32}$$

$$v(0) = v(1) = 0, \tag{9.33}$$

then

$$\begin{aligned} \|(u - u_S)(t)\|_2 &\leq \|(u - \tilde{u}_S)(t)\|_2 + [\|u_S(0) - \tilde{u}_S(0)\|_2^2 \\ &\quad + t[2\gamma\pi^2(1 - \lambda\Lambda^{-1})]^{-1} \sup_{0 \leq s \leq t} \|D_t(u - \tilde{u}_S)(s)\|_2^2]^{1/2}, \qquad t > 0. \end{aligned} \tag{9.34}$$

Proof. By our hypotheses, we have that for all $w \in S$,

$$
\begin{aligned}
(D_t(u - \tilde{u}_S)(t), w)_2 &= (f(u(t)), w)_2 - a(u(t), w) - (D_t\tilde{u}_S(t), w)_2 \\
&= (D_t u_S(t), w)_2 + a(u_S(t), w) - (f(u_S(t)), w)_2 \\
&\quad + (f(\tilde{u}_S(t)), w)_2 - a(\tilde{u}_S(t), w) - (D_t\tilde{u}_S(t), w)_2.
\end{aligned}
\tag{9.35}
$$

Thus,

$$
\begin{aligned}
&(D_t(u - \tilde{u}_S)(t), w)_2 \\
&\quad = (D_t(u_S - \tilde{u}_S)(t), w)_2 + a((u_S - \tilde{u}_S)(t), w) - (f(u_S(t)) - f(\tilde{u}_S(t)), w)_2
\end{aligned}
$$

and, choosing $w \equiv u_S(t) - \tilde{u}_S(t)$, we have

$$
\begin{aligned}
&(D_t(u - \tilde{u}_S)(t), (u_S - \tilde{u}_S)(t))_2 \\
&\quad = (D_t(u_S - \tilde{u}_S)(t), (u_S - \tilde{u}_S)(t))_2 + a((u_S - \tilde{u}_S)(t), (u_S - \tilde{u}_S)(t)) \\
&\quad\quad - (f(u_S(t)) - f(\tilde{u}_S(t)), (u_S - \tilde{u}_S)(t))_2 \\
&\quad \geq \tfrac{1}{2} D_t \| (u_S - \tilde{u}_S)(t) \|_2^2 + (1 - \lambda\Lambda^{-1}) a((u_S - \tilde{u}_S)(t), (u_S - \tilde{u}_S)(t)).
\end{aligned}
$$

But

$$
\begin{aligned}
(D_t(u - \tilde{u}_S)(t), (u_S - \tilde{u}_S)(t))_2 &\leq [4\gamma(1 - \lambda\Lambda^{-1})\pi^2]^{-1} \| D_t(u - \tilde{u}_S)(t) \|_2^2 \\
&\quad + \gamma(1 - \lambda\Lambda^{-1})\pi^2 \| (u_S - \tilde{u}_S)(t) \|_2^2,
\end{aligned}
$$

and hence

$$
\begin{aligned}
\tfrac{1}{2} D_t \| (u_S - \tilde{u}_S)(t) \|_2^2 &\leq [4\gamma(1 - \lambda\Lambda^{-1})\pi^2]^{-1} \| D_t(u - \tilde{u}_S)(t) \|_2^2 \\
&\quad + \gamma(1 - \lambda\Lambda^{-1})\pi^2 \| (u_S - \tilde{u}_S)(t) \|_2^2 \\
&\quad - (1 - \lambda\Lambda^{-1}) a((u_S - \tilde{u}_S)(t), (u_S - \tilde{u}_S)(t)) \\
&\leq [4\gamma(1 - \lambda\Lambda)^{-1})\pi^2]^{-1} \| D_t(u - \tilde{u}_S)(t) \|_2^2.
\end{aligned}
\tag{9.36}
$$

Integrating both sides of the inequality (9.36) from 0 to t, we obtain

$$
\begin{aligned}
\| (u_S - \tilde{u}_S)(t) \|_2^2 &\leq \| u_S(0) - \tilde{u}_S(0) \|_2^2 \\
&\quad + t[2\gamma\pi^2(1 - \lambda\Lambda^{-1})]^{-1} \sup_{0 \leq s \leq t} \| D_t(u - \tilde{u}_S)(s) \|_2^2,
\end{aligned}
\tag{9.37}
$$

and the result follows from the triangle inequality. Q.E.D.

Using the results of Chapter 7 to bound the quantities on the right-hand side of (9.34), we obtain the following results; cf. [9.3] and [9.8] for the technical details.

THEOREM 9.9

If $a(u, v)$ is strongly coercive and $u \in PC^{3,\infty}(I \times (0, \infty))$, then there exists a positive constant, K, such that

$$\|(u - u_L)(t)\|_2 \leq Kh^2, \qquad t > 0, \tag{9.38}$$

where u_L is the semidiscrete Galerkin approximation to u over $L_0(\Delta)$.

THEOREM 9.10

If $a(u, v)$ is strongly coercive and $u \in PC^{5,\infty}(I \times (0, \infty))$, then there exists a positive constant, K, such that

$$\|(u - u_H)(t)\|_2 \leq Kh^4, \qquad t > 0 \tag{9.39}$$

and

$$\|(u - u_S)(t)\|_2 \leq Kh^4, \qquad t > 0, \tag{9.40}$$

where u_H and u_S are the semi-discrete Galerkin approximations to u over $H_0(\Delta)$ and $S_0(\Delta)$, respectively.

Thus we have shown that under suitable hypotheses on the semilinear problem, the semi-discrete Galerkin approximation to u over $L_0(\Delta)$ is second-order accurate with respect to the L^2-norm, while the semi-discrete Galerkin approximations to u over $H_0(\Delta)$ and $S_0(\Delta)$ are both fourth-order accurate with respect to the L^2-norm.

9.3 COMPUTATIONAL CONSIDERATIONS

In this section, we discuss the question of actually solving the systems of ordinary differential equations, (9.8)–(9.9), which we obtain by discretizing the space variables via Galerkin's procedure. We will treat only the special problems studied in Section 9.1. See [9.2], [9.3], [9.7], and [9.8] for the analogous details about more general problems.

We recall that our system is of the form

$$BD_t\boldsymbol{\beta}(t) + A\boldsymbol{\beta}(t) = \mathbf{k}, \qquad t > 0, \tag{9.8}$$

$$B\boldsymbol{\beta}(0) = \mathbf{g}, \tag{9.9}$$

where A and B are symmetric, positive definite matrices. If $B^{1/2}$ is the unique, nonsingular square root of B, i.e., $(B^{1/2})^2 = A$, $\boldsymbol{\gamma}(t) \equiv B^{1/2}\boldsymbol{\beta}(t)$, and $E \equiv B^{-1/2}AB^{-1/2}$, then (9.8)–(9.9) can be rewritten as

$$D_t\boldsymbol{\gamma}(t) = -E\boldsymbol{\gamma}(t) + \mathbf{c}, \qquad t > 0, \tag{9.41}$$

$$\boldsymbol{\gamma}(0) = \mathbf{h}, \tag{9.42}$$

where $\mathbf{c} \equiv B^{1/2}B^{-1}\mathbf{k} = B^{-1/2}\mathbf{k}$ and $\mathbf{h} \equiv B^{-1/2}\mathbf{g}$. The solution of (9.41)–(9.42) is given by

$$\boldsymbol{\gamma}(t) = e^{-tE}\mathbf{h} + (I - e^{-tE})E^{-1}\mathbf{c} \qquad t \geq 0, \tag{9.43}$$

or equivalently

$$\begin{aligned} \boldsymbol{\gamma}(t_0 + \Delta t) &= e^{-\Delta t E}\boldsymbol{\gamma}(t_0) + (I - e^{-\Delta t E})E^{-1}\mathbf{c} \\ &= E^{-1}\mathbf{c} + e^{-\Delta t E}(\boldsymbol{\gamma}(t_0) - E^{-1}\mathbf{c}). \end{aligned} \tag{9.44}$$

The method that we suggest for computing $\boldsymbol{\gamma}(t)$ is to discretize the continuous time variable $t \in [0, \infty]$ into $\{t_j \equiv j\Delta t \,|\, j = 0, 1, 2, \ldots\}$ and to replace (9.41) by a finite difference equation. If we think of $\boldsymbol{\gamma}_j$ as an approximation to $\boldsymbol{\gamma}(t_j)$, $j = 0, 1, 2, \ldots$, then three well-known finite difference approximations to (9.41) are the *forward difference approximation*

$$(\boldsymbol{\gamma}_{j+1} - \boldsymbol{\gamma}_j)(\Delta t)^{-1} = -E\boldsymbol{\gamma}_j + \mathbf{c}, \qquad j = 0, 1, 2, \ldots, \tag{9.45}$$

or equivalently

$$\boldsymbol{\gamma}_{j+1} = (I - \Delta t E)\boldsymbol{\gamma}_j + \Delta t\mathbf{c}, \qquad j = 0, 1, 2, \ldots, \tag{9.46}$$

the *backward difference approximation*

$$(\boldsymbol{\gamma}_{j+1} - \boldsymbol{\gamma}_j)(\Delta t)^{-1} = -E\boldsymbol{\gamma}_{j+1} + \mathbf{c}, \qquad j = 0, 1, 2, \ldots., \tag{9.47}$$

or equivalently

$$\boldsymbol{\gamma}_{j+1} = (I + \Delta t E)^{-1}\boldsymbol{\gamma}_j + \Delta t(I + \Delta t E)^{-1}\mathbf{c}, \qquad j = 0, 1, 2, \ldots, \tag{9.48}$$

and the *Crank–Nicholson approximation*

$$(\boldsymbol{\gamma}_{j+1} - \boldsymbol{\gamma}_j)(\Delta t)^{-1} = -\frac{E}{2}(\boldsymbol{\gamma}_{j+1} + \boldsymbol{\gamma}_j) + \mathbf{c}, \qquad j = 0, 1, 2, \ldots, \tag{9.49}$$

or equivalently

$$\boldsymbol{\gamma}_{j+1} = (I + \tfrac{1}{2}\Delta t E)^{-1}(I - \tfrac{1}{2}\Delta t E)\boldsymbol{\gamma}_j + \Delta t(I + \tfrac{1}{2}\Delta t E)^{-1}\mathbf{c}. \tag{9.50}$$

Comparing (9.46), (9.48), and (9.50) with (9.44), we see that these three difference approximations give rise to particular rational matrix approximations to the matrix $e^{-\Delta t E}$.

Following [9.9], these approximations may be generalized and studied from the viewpoint of the Padé table for e^{-z}. The Padé table for e^{-z} is a double entry table of rational approximations, $R_{p,q}(z)$, such that

$$R_{p,q}(z) \equiv n_{p,q}(z)[d_{p,q}(z)]^{-1} = e^{-z} + O(|z|^r) \qquad \text{as } |z| \longrightarrow 0, \tag{9.51}$$

(where $n_{p,q}(z)$ and $d_{p,q}(z)$ are polynomials of degree q and p respectively) gives the best approximation for e^{-z} near $z = 0$. Except for a multiplicative factor, these polynomials are uniquely determined and $r = p + q + 1$. In fact, it turns out that

$$n_{p,q}(z) = \sum_{k=0}^{q} (p + q - k)![(p + q)!k!(q - k)!]^{-1}(-z)^k, \tag{9.52}$$

$$d_{p,q}(z) = \sum_{k=0}^{p} (p + q - k)!p![(p + q)!k!(p - k)!]z^k, \tag{9.53}$$

and

$$|R_{p,q}(z)| \leq p!q!z^{p+q+1}e^{-z}[(p + q)!(p + q + 1)!d_{p,q}(z)]^{-1}, \tag{9.54}$$

for all real z. Clearly, $R_{0,1}(z) = 1 - z$, $R_{1,0}(z) = (1 + z)^{-1}$, and $R_{1,1}(z) = (1 - \frac{1}{2}z)(1 + \frac{1}{2}z)^{-1}$.

To define our class of difference approximations, let $R_{p,q}(\Delta tE)$ denote the matrix Padé approximation to $e^{-\Delta tE}$ and

$$\boldsymbol{\gamma}_{j+1}(p, q) = E^{-1}\mathbf{c} + R_{p,q}(\Delta tE)(\boldsymbol{\gamma}_j(p, q) - E^{-1}\mathbf{c}), \qquad j = 0, 1, 2, \ldots \tag{9.55}$$

$$\boldsymbol{\gamma}_0(p, q) = \mathbf{h}. \tag{9.56}$$

We now state and prove a result giving the necessary and sufficient conditions for stability in the L^2-norm for this class of difference approximations coupled with the Galerkin approximations.

THEOREM 9.11

Let S be the finite-dimensional subspace of $PC_0^{1,2}(I)$ spanned by the basis functions $\{B_i(x)\}_{i=1}^n$, which are normalized so that $\mathbf{x}^T\mathbf{x} \leq \mathbf{x}^TB\mathbf{x} \leq \Lambda\mathbf{x}^T\mathbf{x}$ for all $\mathbf{x} \in R^n$. If

$$u_S^{p,q}(t_j) = \sum_{i=1}^{n} \beta_i^{p,q}(j)B_i(x),$$

where $\boldsymbol{\beta}^{p,q}(j) \equiv B^{-1/2}\boldsymbol{\gamma}_j(p, q)$, then

$$\begin{aligned} \|u_S^{p,q}(t_j)\|_2 \leq \operatorname{cond}(B)\{&\gamma^{-1}\|f\|_2 \\ &+ |R_{p,q}^j(\Delta tE)|_2(\|u_0\|_2 + \gamma^{-1}\|f\|_2)\}, \qquad j = 0, 1, 2, \ldots, \end{aligned} \tag{9.57}$$

where $|\mathbf{x}|_2 \equiv \left(\sum_{i=1}^{n} x_i^2\right)^{1/2}$ for all $\mathbf{x} \in R^n$ and

$$|M|_2 \equiv \sup\{|M\mathbf{x}|_2|\mathbf{x}|_2^{-1} \mid \mathbf{x} \in R^n, \mathbf{x} \neq \mathbf{0}\}$$

for all $n \times n$ matrices.

Proof. Using our hypotheses, we have for all $j = 0, 1, 2, \ldots,$

$$\begin{aligned} \| u_S^{p,q}(t_j) \|_2 &\leq \Lambda^{1/2} | \boldsymbol{\beta}^{p,q}(j) |_2 \\ &\leq \Lambda^{1/2} | B^{-1/2} |_2 | \boldsymbol{\gamma}_j(p, q) |_2 \\ &\leq \Lambda^{1/2} | B^{-1/2} |_2 \{ | E^{-1}\mathbf{c} |_2 + | R_{p,q}^j(\Delta t E) |_2 | \mathbf{h} - E^{-1}\mathbf{c} |_2 \} \\ &\leq \Lambda^{1/2} | B^{-1/2} |_2 \{ | B^{1/2} A^{-1}\mathbf{k} |_2 \\ &\quad + | R_{p,q}^j(\Delta t E) |_2 | B^{-1/2}\mathbf{g} - B^{1/2} A^{-1}\mathbf{k} |_2 \} \\ &\leq \Lambda^{1/2} | B^{-1/2} |_2 | B^{1/2} |_2 \{ | A^{-1}\mathbf{k} |_2 \\ &\quad + | R_{p,q}^j(\Delta t E) |_2 (| B^{-1}\mathbf{g} |_2 + | A^{-1}\mathbf{k} |_2) \}. \end{aligned} \tag{9.58}$$

Moreover, since $\gamma \leq (\mathbf{x}^T A \mathbf{x})(\mathbf{x}^T B \mathbf{x})^{-1}$ for all $\mathbf{x} \in R^n$, we have

$$\gamma \mathbf{x}^T \mathbf{x} \leq \mathbf{x}^T A \mathbf{x}, \qquad \text{for all } \mathbf{x} \in R^n. \tag{9.59}$$

Using this inequality to bound the right-hand side of (9.58), we obtain for all $j = 0, 1, 2, \ldots$

$$\begin{aligned} \| u_S^{p,q}(t_j) \|_2 &\leq \Lambda \{ \gamma^{-1} | \mathbf{k} |_2 + | R_{p,q}^j(\Delta t E) |_2 (| \mathbf{g} |_2 + \gamma^{-1} | \mathbf{k} |_2) \} \\ &\leq \Lambda \{ \gamma^{-1} \| f \|_2 + | R_{p,q}^j(\Delta t E) |_2 (\| u_0 \|_2 + \gamma^{-1} \| f \|_2) \}. \end{aligned} \tag{9.60}$$

Q.E.D.

COROLLARY 1

Let C be a set of finite dimensional subspaces, $S \equiv \text{span}\,\{B_i(x)\}_{i=1}^{\dim S}$, of $PC_0^{1,2}(I)$ such that

$$\mathbf{x}^T \mathbf{x} \leq \mathbf{x}^T B \mathbf{x} \leq \Lambda \mathbf{x}^T \mathbf{x}, \qquad \text{for all } \mathbf{x} \in R^{\dim S}, \tag{9.61}$$

where Λ is a positive constant independent of $S \in C$. If

$$\tau_{p,q} \equiv \sup \{ t \geq 0 \mid n_{p,q}(z)[d_{p,q}(z)]^{-1} \leq 1 \qquad \text{for } 0 \leq z \leq t \},$$

then

$$\| u_S^{p,q}(t_j) \|_2 \leq \Lambda (\| u_0 \|_2 + 2\gamma^{-1} \| f \|_2), \qquad \text{for all } j = 0, 1, 2, \ldots, \quad S \in C, \tag{9.62}$$

and

$$0 \leq \Delta t \leq \tau_{p,q} \rho^{-1}(E), \tag{9.63}$$

where $\rho(E)$ denotes the spectral radius of E (cf. [9.9]).

Proof. Since E is symmetric, we have that if (9.62) holds then

$$| R_{p,q}^j(\Delta t E) |_2 = [\rho(R_{p,q}(\Delta t E))]^j \leq 1$$

(cf. [9.9]). The result then follows from (9.57). Q.E.D.

Inequality (9.62) states that the discretization procedure of Corollary 1 is *strongly stable.*

COROLLARY 2

If $p \geq q$ and the hypotheses of Corollary 1 hold, then

$$\| u_S^{p,q}(t_j) \|_2 \leq \Lambda (\| u_0 \|_2 + 2\gamma^{-1} \| f \|_2), \tag{9.64}$$

for all $j = 0, 1, 2, \ldots$, all $\Delta t > 0$, and all $S \in C$.

Thus, we have shown that under the appropriate hypotheses the backwards difference and Crank–Nicholson difference approximations coupled with the Galerkin method are strongly stable for all $\Delta t > 0$. The spaces $L_0(\Delta)$, $H_0(\Delta)$, and $S_0(\Delta)$ with the basis functions given in Chapter 7 satisfy these hypotheses.

We now show how to obtain a priori error bounds for the important special case of the Crank–Nicholson–Galerkin approximation.

THEOREM 9.12

Let S be a finite-dimensional subspace of $PC_0^{1,2}(I)$ and $\tilde{u}_S(t_j)$, $j = 0, 1, 2, \ldots$, be the orthogonal projection of $u(x, t)$ onto S with respect to the inner-product $a(u, v)$. If $u \in C^3(I \times [0, \infty])$, then there exists a positive constant, K, such that

$$\begin{aligned} \| u(x, t_j) - u_S^{1,1}(t_j) \|_2 \leq{}& \| u(x, t_j) - \tilde{u}_S(t_j) \|_2 \\ &+ [2\gamma^{-1}\pi^{-2} \sup_{1 \leq k \leq j} (\| D_+(u(x, t_k) - \tilde{u}_S(t_k)) \|_2^2 \\ &+ K(\Delta t)^2 \sup_{0 \leq t \leq (k+1)\Delta t} \| D_t^3 u \|_2^2) \\ &+ \| u_S^{1,1}(0) - \tilde{u}_S(0) \|_2^2]^{1/2}, \qquad j = 0, 1, 2, \ldots, \end{aligned} \tag{9.65}$$

where D_+ is the forward difference quotient operator in the time variable, i.e.,

$$D_+ z_j \equiv (z_{j+1} - z_j)(\Delta t)^{-1}.$$

Proof. We let $\delta_j \equiv u(x, t_j) - \tilde{u}_S(t_j)$ and $\epsilon_j \equiv u_S^{1,1}(t_j) - \tilde{u}_S(t_j)$, for all $j = 0, 1, 2, \ldots$. Then by Taylor's theorem applied to $u(x, t)$, we have

$$\begin{aligned} (D_+\delta_j, w)_2 = (f, w)_2 - a(\tfrac{1}{2}(u(t_{j+1}) + u(t_j)), w) \\ - (D_+\tilde{u}_S(t_j), w)_2 + (e_j, w)_2, \end{aligned} \tag{9.66}$$

for all $w \in S$ and all $j = 0, 1, 2, \ldots$, where

$$\| e_j \|_2 \leq K(\Delta t)^2 \sup_{0 \leq t \leq (j+1)\Delta t} \| D_t^3 u(x, t) \|_2.$$

Thus,

$$(9.67)\quad \begin{aligned}(D_+\delta_j, w)_2 = (D_+u_S^{1,1}(t_j), w)_2 + a(\tfrac{1}{2}(u_S^{1,1}(t_{j+1}) + u_S^{1,1}(t_j)), w)\\ - a(\tfrac{1}{2}(\tilde{u}_S(t_{j+1}) + \tilde{u}_S(t_j)), w) - (D_+\tilde{u}_S(t_j), w)_2 + (e_j, w)_2,\end{aligned}$$

for all $w \in S$ and all $j = 0, 1, 2, \ldots$. Setting $w \equiv \epsilon_j + \epsilon_{j+1}$ in (9.67), we have

$$(9.68)\quad \begin{aligned}(D_+\delta_j, \epsilon_j + \epsilon_{j+1})_2 &= (D_+\epsilon_j, \epsilon_j + \epsilon_{j+1})_2 + \tfrac{1}{2}a(\epsilon_j + \epsilon_{j+1}, \epsilon_j + \epsilon_{j+1})\\ &\quad + (e_j, \epsilon_j + \epsilon_{j+1})_2\\ &\geq \tfrac{1}{2}D_+ \|\epsilon_j\|_2^2 + \tfrac{1}{2}\gamma\pi^2 \|\epsilon_j + \epsilon_{j+1}\|_2^2 - \gamma^{-1}\pi^{-2} \|e_j\|_2^2\\ &\quad - \tfrac{1}{4}\gamma\pi^2 \|\epsilon_j + \epsilon_{j+1}\|_2^2, \qquad j = 0, 1, 2, \ldots,\end{aligned}$$

where we have used the fact that $ab \geq -\eta^{-1}a^2 - \frac{1}{4}\eta b^2$ for all $\eta > 0$ and all a and b.

Thus, we have

$$(9.69)\quad \begin{aligned}\tfrac{1}{2}D_+ \|\epsilon_j\|_2^2 &\leq (D_+\delta_j, \epsilon_j + \epsilon_{j+1})_2 + \gamma^{-1}\pi^{-2} \|e_j\|_2^2 - \tfrac{1}{4}\gamma\pi^2 \|\epsilon_j + \epsilon_{j+1}\|_2^2\\ &\leq \gamma^{-1}\pi^2 \|D_+\delta_j\|_2^2 + \gamma^{-1}\pi^2 \|e_j\|_2^2, \qquad j = 0, 1, 2, \ldots,\end{aligned}$$

and it follows that

$$(9.70)\quad \|\epsilon_j\|_2^2 \leq 2\gamma^{-1}\pi^{-2} \sup_{1\leq k\leq j} (\|D_+\delta_k\|_2^2 + \|e_k\|_2^2) + \|e_0\|_2, \qquad j = 1, 2, \ldots.$$

Inequality (9.65) follows from (9.70) and the triangle inequality. Q.E.D.

Combining Theorem 9.12 with the results of Chapter 7, we find that under the appropriate hypotheses

$$(9.71)\quad \|u - u_L^{1,1}\|_2 = O(h^2 + (\Delta t)^2) \qquad \text{as } h \text{ and } \Delta t \longrightarrow 0,$$

$$(9.72)\quad \|u - u_H^{1,1}\|_2 = O(h^4 + (\Delta t)^2) \qquad \text{as } h \text{ and } \Delta t \longrightarrow 0,$$

and

$$(9.73)\quad \|u - u_S^{1,1}\|_2 = O(h^4 + (\Delta t)^2) \qquad \text{as } h \text{ and } \Delta t \longrightarrow 0.$$

The estimates (9.71) and (9.72) suggest the use of Richardson extrapolation with respect to the time difference scheme (cf. [9.4]), and indeed it is possible to show that in these two cases one extrapolation yields an approximation which is fourth-order accurate in time as well as in space.

EXERCISES FOR CHAPTER 9

(9.1) Prove analogues of the results of this chapter for the spaces

$$S_0(2m - 1, \Delta, z) \equiv \{\phi \in S(2m - 1, \Delta, z) \mid \phi(0) = \phi(1) = 0\}$$

defined in Exercise (4.12).

(9.2) Derive bounds for the interval of stability for the forward difference approximation coupled with the semi-discrete Galerkin method over $L_0(\Delta)$, i.e., derive bounds for $\tau_{0,1}\rho^{-1}(E)$.

(9.3) Prove rigorously that one recursion of Richardson extrapolation applied to the Crank–Nicholson difference approximation to the semi-discrete Galerkin equations over $H_0(\Delta)$ and $S_0(\Delta)$ yields a fourth-order accurate approximation with respect to the L^2-norm.

REFERENCES FOR CHAPTER 9

[9.1] BRAUER, F., and J. NOHEL, *Ordinary Differential Equations.* W. A. Benjamin, Inc., New York (1966).

[9.2] DENDY, J., *Penalty Galerkin methods for partial differential equations.* Ph.D. Dissertation, Rice University (June, 1971).

[9.3] DOUGLAS, J., and T. DUPONT, Galerkin methods for parabolic equations. *SIAM J. Numer. Anal.* **7**, 575–626 (1970).

[9.4] ISAACSON, E., and H. B. KELLER, *Analysis of Numerical Methods.* John Wiley & Sons, Inc., New York (1966).

[9.5] LIONS, J. L., *Équationes Différentielles Opérationelles et Problemes aux Limites.* Springer-Verlag, Berlin (1961).

[9.6] PRICE, H. S., and R. S. VARGA, Error bounds for semi-discrete Galerkin approximations of parabolic problems with applications to petroleum reservoir mechanics. *Numerical Solution of Field Problems in Continuum Physics* 74–95, AMS, Providence (1970).

[9.7] SWARTZ, B., and B. WENDROFF, Generalized finite difference schemes. *Math. of Comp.* **23**, 37–49 (1969).

[9.8] WHEELER, M. F., *A priori L^2 error estimates for Galerkin approximations to parabolic differential equations.* Ph.D. Dissertation, Rice University (June, 1971).

[9.9] VARGA, R. S., *Matrix Iterative Analysis.* Prentice-Hall, Inc., Englewood Cliffs, N. J. (1962).

10 THE RITZ PROCEDURE FOR AN OPTIMAL CONTROL PROBLEM

10.1 FORMULATION OF THE PROCEDURE

In this chapter, we consider a variational procedure for approximating the solution of the "state regulator problem" in optimal control. Following Borsage and Johnson (cf. [10.2], [10.3], [10.4], and [10.6]), we consider the Lagrange formulation of the problem and show that the Lagrange multiplier can be characterized as the solution of the variational problem of minimizing a quadratic, positive definite functional, F, over an appropriate function space, Φ_0^n.

We obtain approximate solutions by using Ritz's idea of minimizing F over finite dimensional subspaces of Φ_0^n and derive general a priori error bounds for this procedure in terms of approximation theory. Finally we apply these results to obtain asymptotic error bounds for the subspaces which we have previously considered.

We let $Q(t)$ and $R(t)$ be respectively an $n \times n$ symmetric, positive definite matrix and an $r \times r$ symmetric, positive definite matrix, both of which are continuous functions of $t \in I$. For each $k \geq 1$, we let

$$\Phi^k \equiv \mathop{\times}_{i=1}^{k} [PC^{1,2}(I)]_i.$$

The state regulator problem in optimal control is to find $\mathbf{u}^* \in \Phi^r$ and $\mathbf{x}^* \in \Phi^n$ which minimize

$$J[\mathbf{u}, \mathbf{x}] \equiv \tfrac{1}{2} \int_0^1 \{(\mathbf{x}(t), Q(t)\mathbf{x}(t)) + (\mathbf{u}(t), R(t)\mathbf{u}(t))\}\, dt \tag{10.1}$$

over all $\mathbf{u} \in \Phi^r$, where $\mathbf{x}(t)$ is given by

$$D_t\mathbf{x}(t) = A(t)\mathbf{x}(t) + B(t)\mathbf{u}(t), \qquad 0 < t < 1, \tag{10.2}$$

and

$$\mathbf{x}(0) = \mathbf{x}_0, \tag{10.3}$$

$$|\mathbf{y}|_2^2 \equiv (\mathbf{y}, \mathbf{y}) \equiv \sum_{i=1}^{n} y_i^2, \qquad \text{for all } \mathbf{y} \in R^n,$$

$$|\mathbf{z}|_2^2 \equiv (\mathbf{z}, \mathbf{z}) \equiv \sum_{i=1}^{r} z_i^2, \qquad \text{for all } \mathbf{z} \in R^r,$$

$A(t)$ is an $n \times n$ matrix, and $B(t)$ is an $n \times r$ matrix, both of which have entries which are bounded, piecewise continuous functions of $t \in I$.

Using standard arguments in the calculus of variations (cf. [10.1]), we can show that the state regulator problem is equivalent to the variational problem of finding $\boldsymbol{\lambda}^* \in \Phi^n$ which minimizes

$$\begin{aligned} -L[\mathbf{u}, \mathbf{x}; \boldsymbol{\lambda}, \boldsymbol{\gamma}] &\equiv J[\mathbf{u}, \mathbf{x}] \\ &+ \int_0^1 (\boldsymbol{\lambda}(t), -D_t\mathbf{x}(t) + A(t)\mathbf{x}(t) + B(t)\mathbf{u}(t))\, dt + (\boldsymbol{\gamma}, \mathbf{x}(0) - \mathbf{x}_0), \end{aligned} \tag{10.4}$$

subject to the constraint that

$$\boldsymbol{\lambda}(1) = \mathbf{0}, \tag{10.5}$$

where $\boldsymbol{\gamma}$, $\mathbf{u}(t)$, and $\mathbf{x}(t)$ are given by

$$\boldsymbol{\gamma} = -\boldsymbol{\lambda}(0), \tag{10.6}$$

$$\mathbf{u}(t) = -R^{-1}(t)B^T(t)\boldsymbol{\lambda}(t), \qquad \text{for all } t \in I, \tag{10.7}$$

and

$$\mathbf{x}(t) = -Q^{-1}(t)(D_t\boldsymbol{\lambda}(t) + A^T(t)\boldsymbol{\lambda}(t)), \qquad \text{for all } t \in I. \tag{10.8}$$

Using the characterizations (10.6), (10.7), and (10.8), we can express $L[\mathbf{u}, \mathbf{x}; \boldsymbol{\lambda}, \boldsymbol{\gamma}]$ in terms of $\boldsymbol{\lambda}$ only. In fact, we have

$$\begin{aligned} -L[\mathbf{u}, \mathbf{x}; \boldsymbol{\lambda}, \boldsymbol{\gamma}] &= -J[\mathbf{u}, \mathbf{x}] + (\boldsymbol{\lambda}(t), \mathbf{x}(t))\big|_0^1 \\ &\quad - \int_0^1 (D_t\boldsymbol{\lambda} + A^T\boldsymbol{\lambda}, \mathbf{x})\, dt - \int_0^1 (B^T\boldsymbol{\lambda}, \mathbf{u})\, dt \\ &\quad + (\boldsymbol{\lambda}(0), \mathbf{x}(0) - \mathbf{x}_0) \\ &= J[\mathbf{u}, \mathbf{x}] - (\boldsymbol{\lambda}(0), \mathbf{x}_0). \end{aligned}$$

But

$$\begin{aligned} \tfrac{1}{2} \int_0^1 (\mathbf{u}, R\mathbf{u})\, dt &= \tfrac{1}{2} \int_0^1 (R^{-1}B^T\boldsymbol{\lambda}, RR^{-1}B^T\boldsymbol{\lambda}) \\ &= \tfrac{1}{2} \int_0^1 (BR^{-1}B^T\boldsymbol{\lambda}, \boldsymbol{\lambda})\, dt, \end{aligned}$$

and

$$\begin{aligned}
\tfrac{1}{2}\int_0^1 (\mathbf{x}, Q\mathbf{x})\,dt &= \tfrac{1}{2}\int_0^1 (Q^{-1}A^T\boldsymbol{\lambda} + Q^{-1}D_t\boldsymbol{\lambda}, A^T\boldsymbol{\lambda} + D_t\boldsymbol{\lambda})\,dt \\
&= \tfrac{1}{2}\int_0^1 (Q^{-1}D_t\boldsymbol{\lambda}, D_t\boldsymbol{\lambda})\,dt + \tfrac{1}{2}\int_0^1 (Q^{-1}A^T\boldsymbol{\lambda}, A^T\boldsymbol{\lambda})\,dt \\
&\quad + \tfrac{1}{2}\int_0^1 (Q^{-1}A^T\boldsymbol{\lambda}, D_t\boldsymbol{\lambda})\,dt + \tfrac{1}{2}\int_0^1 (Q^{-1}D_t\boldsymbol{\lambda}, A^T\boldsymbol{\lambda})\,dt \\
&= \tfrac{1}{2}\int_0^1 (Q^{-1}D_t\boldsymbol{\lambda}, D_t\boldsymbol{\lambda})\,dt + \tfrac{1}{2}\int_0^1 (AQ^{-1}A^T\boldsymbol{\lambda}, \boldsymbol{\lambda})\,dt \\
&\quad + \int_0^1 (AQ^{-1}D_t\boldsymbol{\lambda}, \boldsymbol{\lambda})dt.
\end{aligned}$$

Thus,

$$\begin{aligned}
F[\boldsymbol{\lambda}] &\equiv -L[\mathbf{u}, \mathbf{x}; \boldsymbol{\lambda}, \boldsymbol{\gamma}] \\
&= \tfrac{1}{2}\int_0^1 (Q^{-1}D_t\boldsymbol{\lambda}, D_t\boldsymbol{\lambda})\,dt + \tfrac{1}{2}\int_0^1 (AQ^{-1}A^T\boldsymbol{\lambda}, \boldsymbol{\lambda})\,dt \\
&\quad + \tfrac{1}{2}\int_0^1 (BR^{-1}B^T\boldsymbol{\lambda}, \boldsymbol{\lambda})\,dt + \int_0^1 (AQ^{-1}D_t\boldsymbol{\lambda}, \boldsymbol{\lambda})\,dt - (\boldsymbol{\lambda}(0), \mathbf{x}_0).
\end{aligned}$$

If we define

$$\begin{aligned}
[\boldsymbol{\lambda}, \boldsymbol{\eta}] &\equiv \int_0^1 (Q^{-1}D_t\boldsymbol{\lambda}, D_t\boldsymbol{\eta})\,dt + \int_0^1 (AQ^{-1}A^T\boldsymbol{\lambda}, \boldsymbol{\eta})\,dt \\
&\quad + \int_0^1 (BR^{-1}B^T\boldsymbol{\lambda}, \boldsymbol{\eta})\,dt + \int_0^1 \{(AQ^{-1}D_t\boldsymbol{\lambda}, \boldsymbol{\eta}) \\
&\quad + (AQ^{-1}D_t\boldsymbol{\eta}, \boldsymbol{\lambda})\}\,dt,
\end{aligned} \tag{10.9}$$

for all $\boldsymbol{\lambda}$ and $\boldsymbol{\eta}$ in Φ^n, then

$$F[\boldsymbol{\lambda}] = \tfrac{1}{2}[\boldsymbol{\lambda}, \boldsymbol{\lambda}] - (\boldsymbol{\lambda}(0), \mathbf{x}_0). \tag{10.10}$$

If we use the notation that for any $r \times p$ matrix M,

$$|M|_2 \equiv \max\{|Mx|_2 \mid \mathbf{x} \in R^p \quad \text{and} \quad |\mathbf{x}|_2 = 1\},$$

we may prove the following characterization result.

THEOREM 10.1

The optimal Lagrange multiplier exists and is the unique solution in

$$\Phi_0^n \equiv \{\boldsymbol{\phi} \in \Phi^n \mid \boldsymbol{\phi}(1) = \mathbf{0}\}$$

of the generalized Euler equation

$$[\boldsymbol{\lambda}, \boldsymbol{\eta}] = (\boldsymbol{\eta}(0), \mathbf{x}_0), \qquad \text{for all } \boldsymbol{\eta} \in \Phi_0^n. \tag{10.11}$$

Moreover, $[\boldsymbol{\lambda}, \boldsymbol{\eta}]$ is symmetric,

$$\text{(10.12)} \quad \|D_t\boldsymbol{\lambda}\|_2^2 \equiv \int_0^1 (D_t\boldsymbol{\lambda}, D_t\boldsymbol{\lambda})\, dt \leq 2\xi_Q^{-1}[\|Q\|_2 + \rho\|A^T\|_2]^2[\boldsymbol{\lambda}, \boldsymbol{\lambda}],$$

and

$$\text{(10.13)} \qquad \|\boldsymbol{\lambda}\|_2^2 \leq 2\xi_Q^{-1}\rho^2[\boldsymbol{\lambda}, \boldsymbol{\lambda}],$$

where $\xi_Q \equiv \min_{t \in I} \{\xi(t) \mid \xi(t)$ is an eigenvalue of $Q(t)\}$,

$$\|Q\|_2^2 \equiv \int_0^1 |Q(t)|_2^2\, dt, \; \|A^T\|_2^2 \equiv \int_0^1 |A^T(t)|_2^2\, dt,$$

$$\|A^T\|_\infty \equiv \max_{t \in I} |A^T(t)|_2, \quad \text{and} \quad \rho \equiv \|Q\|_2(2\|A^T\|_\infty)^{-1/2}e^{\|A^T\|_\infty}.$$

Proof. The existence part of the theorem is a standard result in optimal control theory; cf. [10.1]. If $\boldsymbol{\eta} \in \Phi_0^n$ and $\alpha \in R$, we have $F[\boldsymbol{\lambda}^* + \alpha\boldsymbol{\eta}] \geq F[\boldsymbol{\lambda}^*]$, with equality if and only if $\alpha = 0$. Hence, we must have $(\partial F/\partial\alpha)[\boldsymbol{\lambda}^*] = 0$, and this implies that (10.11) holds.

Clearly, $[\boldsymbol{\lambda}, \boldsymbol{\eta}]$ is symmetric in $\boldsymbol{\lambda}$ and $\boldsymbol{\eta}$ and

$$[\boldsymbol{\lambda}, \boldsymbol{\lambda}] = \tfrac{1}{2}\int_0^1 (\mathbf{u}, R\mathbf{u})\, dt + \tfrac{1}{2}\int_0^1 (\mathbf{x}, Q\mathbf{x})\, dt$$
$$\geq \tfrac{1}{2}\int_0^1 (\mathbf{x}, Q\mathbf{x})\, dt \geq \tfrac{1}{2}\xi_Q \int_0^1 (\mathbf{x}, \mathbf{x})\, dt,$$

where $\mathbf{u}$ and $\mathbf{x}$ are given by (10.7) and (10.8). From (10.8), we have

$$|\boldsymbol{\lambda}(t)|_2 \equiv (\boldsymbol{\lambda}(t), \quad \boldsymbol{\lambda}(t))^{1/2} \leq \|Q\|_2\|\mathbf{x}\|_2 + \int_t^1 |A^T(s)|_2 |\boldsymbol{\lambda}(s)|_2\, dt,$$

and using Gronwall's Inequality (cf. [10.5]), we obtain the inequality

$$\|\boldsymbol{\lambda}\|_2 \leq \|Q\|_2(2\|A^T\|_\infty)^{-1/2}e^{\|A^T\|_\infty}\|\mathbf{x}\|_2 \equiv \rho\|\mathbf{x}\|_2.$$

Thus, we have $[\boldsymbol{\lambda}, \boldsymbol{\lambda}] \geq \frac{1}{2}\xi_Q\rho^{-2}\|\boldsymbol{\lambda}\|_2^2$, which proves (10.13), and (10.12) follows by combining (10.8) and (10.13).

Finally, if $\boldsymbol{\lambda}$ and $\boldsymbol{\eta}$ both satisfy (10.11), then

$$0 = [\boldsymbol{\lambda} - \boldsymbol{\mu}, \boldsymbol{\lambda} - \boldsymbol{\mu}] \geq \tfrac{1}{2}\xi_Q\rho^{-2}\|\boldsymbol{\lambda} - \boldsymbol{\mu}\|_2^2$$

and $\boldsymbol{\lambda} = \boldsymbol{\mu}$, which proves the uniqueness result. Q.E.D.

To define the Ritz approximation procedure, we let S be any finite-dimensional subspace of Φ_0^n and we find an approximation, $\boldsymbol{\lambda}_S$, to $\boldsymbol{\lambda}^*$ by minimizing F over S and an approximation, $\mathbf{u}_S$, to $\mathbf{u}^*$ via equation (10.7). When we apply the computed control, we obtain the state $\mathbf{x}_S$ determined by

(10.2). It is important to note that $\mathbf{x}_S$ is *not* the state which we can compute via equation (10.8).

We now show that the Ritz procedure is well-defined.

THEOREM 10.2

There exists a unique $\boldsymbol{\lambda}_S \in S$ which minimizes F over S.

Proof. Let $\{\mathbf{B}_i(t)\}_{i=1}^m$ be a basis for S. Considering

$$F[\sum_{i=1}^m \beta_i \mathbf{B}_i] = \tfrac{1}{2}[\sum_{i=1}^m \beta_i \mathbf{B}_i, \sum_{i=1}^m \beta_i \mathbf{B}_i] - (\sum_{i=1}^m \beta_i \mathbf{B}_i(0), \mathbf{x}_0)$$

as a function of $\boldsymbol{\beta} \in R^m$, it is clear that F is twice continuously differentiable and hence F has a minimum at $\boldsymbol{\beta}^*$ if and only if

$$\text{(10.14)} \qquad \frac{\partial F}{\partial \beta_i}[\boldsymbol{\beta}^*] = 0, \qquad \text{for all } 1 \le i \le m,$$

and the Hessian matrix of F, $H \equiv [\partial^2 F/\partial x_i \partial x_j]$, is positive definite. Calculating the equations (10.14), we obtain

$$\text{(10.15)} \qquad \frac{\partial F}{\partial \beta_i}[\boldsymbol{\beta}^*] = \sum_{j=1}^m \beta_j^*[\mathbf{B}_i, \mathbf{B}_j] - (\mathbf{B}_i(0), \mathbf{x}_0), \qquad 1 \le i \le m,$$

or

$$\text{(10.16)} \qquad A\boldsymbol{\beta}^* = \mathbf{k},$$

where

$$\text{(10.17)} \qquad A \equiv [[\mathbf{B}_i, \mathbf{B}_j]]$$

and

$$\text{(10.18)} \qquad \mathbf{k} \equiv [(\mathbf{B}_i(0), \mathbf{x}_0)].$$

Clearly, A is symmetric and positive definite. In fact, if $\boldsymbol{\beta} \neq \mathbf{0}$, then from (10.13) we have

$$\boldsymbol{\beta}^T A \boldsymbol{\beta} = \left[\sum_{i=1}^m \beta_i \mathbf{B}_i, \sum_{i=1}^m \beta_i \mathbf{B}_i\right] \ge \tfrac{1}{2}\xi_Q \rho^{-2} \left\| \sum_{i=1}^m \beta_i B_i \right\|_2 > 0.$$

Moreover, it follows from (10.15) that $H = 2A$ and hence $\boldsymbol{\beta}^*$ is the unique minimum of F over R^m. Q.E.D.

10.2 ERROR BOUNDS

In this section, we obtain general error bounds and then apply the results of Chapters 2, 3, and 4 to obtain error bounds for the spaces which we have previously discussed.

THEOREM 10.3

If $\boldsymbol{\lambda}_S$ denotes the Ritz approximation to $\boldsymbol{\lambda}^*$ over S, then

$$|\boldsymbol{\lambda}^* - \boldsymbol{\lambda}_S| \equiv [\boldsymbol{\lambda}^* - \boldsymbol{\lambda}_S, \boldsymbol{\lambda}^* - \boldsymbol{\lambda}_S]^{1/2} = \inf_{\mathbf{w} \in S} |\boldsymbol{\lambda}^* - \mathbf{w}|. \tag{10.19}$$

Proof. If $\mathbf{w} \in S$,

$$F[\mathbf{w}] = \tfrac{1}{2}[\mathbf{w}, \mathbf{w}] - (\mathbf{w}(0), \mathbf{x}_0)$$

and

$$F[\mathbf{w}] - F[\boldsymbol{\lambda}^*] = \tfrac{1}{2}[\mathbf{w}, \mathbf{w}] - \tfrac{1}{2}[\boldsymbol{\lambda}^*, \boldsymbol{\lambda}^*] + (\mathbf{x}_0, \boldsymbol{\lambda}^*(0) - \mathbf{w}(0)).$$

But taking $\boldsymbol{\eta} = \boldsymbol{\lambda}^*$ in (10.11), we see that $[\boldsymbol{\lambda}^*, \boldsymbol{\lambda}^*] = (\mathbf{x}_0, \boldsymbol{\lambda}^*(0))$, and hence that

$$F[\mathbf{w}] - F[\boldsymbol{\lambda}^*] = \tfrac{1}{2}[\mathbf{w}, \mathbf{w}] + \tfrac{1}{2}[\boldsymbol{\lambda}^*, \boldsymbol{\lambda}^*] + (\mathbf{x}_0, - \mathbf{w}(0)).$$

Taking $\boldsymbol{\eta} = \mathbf{w}$ in (10.11), we see that $[\boldsymbol{\lambda}^*, \mathbf{w}] = (\mathbf{x}_0, \mathbf{w}(0))$, and hence that

$$\begin{aligned} F[\mathbf{w}] - F[\boldsymbol{\lambda}^*] &= \tfrac{1}{2}[\mathbf{w}, \mathbf{w}] + \tfrac{1}{2}[\boldsymbol{\lambda}^*, \boldsymbol{\lambda}^*] - [\boldsymbol{\lambda}^*, \mathbf{w}] \\ &= \tfrac{1}{2}[\boldsymbol{\lambda}^* - \mathbf{w}, \boldsymbol{\lambda}^* - \mathbf{w}] = \tfrac{1}{2}|\boldsymbol{\lambda}^* - \mathbf{w}|^2. \end{aligned}$$

Thus,

$$|\boldsymbol{\lambda}^* - \boldsymbol{\lambda}_S|^2 = 2(F[\boldsymbol{\lambda}_S] - F[\boldsymbol{\lambda}^*]) \leq 2(F[\mathbf{w}] - F[\boldsymbol{\lambda}^*]) = |\boldsymbol{\lambda}^* - \mathbf{w}|^2,$$

and we have

$$\inf_{\mathbf{w} \in S} |\boldsymbol{\lambda}^* - \mathbf{w}| \leq |\boldsymbol{\lambda}^* - \boldsymbol{\lambda}_S| \leq \inf_{\mathbf{w} \in S} |\boldsymbol{\lambda}^* - \mathbf{w}|.$$

Q.E.D.

Combining Theorems 10.1 and 10.3, we have the following result.

COROLLARY

If $\boldsymbol{\lambda}_S$ denotes the Ritz approximation to $\boldsymbol{\lambda}^*$ over S, then

$$\|\boldsymbol{\lambda}^* - \boldsymbol{\lambda}_S\|_2 \leq (2\xi_Q^{-1})^{1/2} \rho \inf_{\mathbf{w} \in S} |\boldsymbol{\lambda}^* - \mathbf{w}| \tag{10.20}$$

and

$$\| D_t(\boldsymbol{\lambda}^* - \boldsymbol{\lambda}_S)\|_2 \leq (2\xi_Q^{-1})^{1/2}(\| D_t Q\|_2 + \rho \| A^T \|_2) \inf_{\mathbf{w} \in S} |\boldsymbol{\lambda}^* - \mathbf{w}|, \tag{10.21}$$

where ξ_Q and ρ are defined in Theorem 10.1.

Using this corollary, we may prove the following results.

THEOREM 10.4

If $\mathbf{u}_S(t) \equiv -R^{-1}(t)B^T(t)\boldsymbol{\lambda}_S(t)$, $t \in \mathrm{I}$, is the computed approximation to $\mathbf{u}^*$, then

$$\|\mathbf{u}^* - \mathbf{u}_S\|_2 \leq \| R^{-1}B^T \|_\infty (2\xi_Q^{-1})^{1/2} \rho \inf_{\mathbf{w} \in S} |\boldsymbol{\lambda}^* - \mathbf{w}|, \tag{10.22}$$

where

$$\| R^{-1}B^T \|_\infty \equiv \sup_{t \in I} | R^{-1}(t)B^T(t) |_2$$

and ξ_Q and ρ are defined in Theorem 10.1.

Proof. In fact, we have that $\boldsymbol{\delta}_S(t) \equiv \mathbf{u}^*(t) - \mathbf{u}_S(t)$ satisfies the equation

$$\boldsymbol{\delta}_S(t) = -R^{-1}(t)B^T(t)(\boldsymbol{\lambda}^*(t) - \boldsymbol{\lambda}_S(t)).$$

Hence, (10.22) follows from the inequality

$$\| \boldsymbol{\delta}_S \|_2 = \| R^{-1}B^T(\boldsymbol{\lambda}^* - \boldsymbol{\lambda}_S) \|_2 \leq \| R^{-1}B^T \|_\infty \| \boldsymbol{\lambda}^* - \boldsymbol{\lambda}_S \|_2$$

and (10.20). Q.E.D.

THEOREM 10.5

If $D_t\mathbf{x}_S(t) = A(t)\mathbf{x}_S(t) + B(t)\mathbf{u}_S(t)$, $t \in I$, and $\mathbf{x}_S(0) = \mathbf{x}_0$, then

$$\| \mathbf{x}^* - \mathbf{x}_S \|_2 \leq \boldsymbol{\Gamma} \| R^{-1}B^T \|_\infty (2\xi_Q^{-1})^{1/2} \rho \inf_{\mathbf{w} \in S} | \boldsymbol{\lambda}^* - \mathbf{w} | \tag{10.23}$$

and

$$\begin{aligned} &\| D_t(\mathbf{x}^* - \mathbf{x}_S) \|_2 \\ &\qquad \leq (\boldsymbol{\Gamma} \| A \|_\infty + \| B \|_\infty) \| R^{-1}B^T \|_\infty (2\xi_Q^{-1})^{1/2} \rho \inf_{\mathbf{w} \in S} | \boldsymbol{\lambda}^* - \mathbf{w} |, \end{aligned} \tag{10.24}$$

where $\boldsymbol{\Gamma} \equiv \| B \|_2 e^{\int_0^1 |A(z)|_2 dz}$, $\| A \|_\infty \equiv \sup_{t \in I} | A(t) |_2$, $\| B \|_\infty \equiv \sup_{t \in I} | B(t) |_2$, and ξ_Q and ρ are defined in Theorem 10.1.

Proof. Letting $\boldsymbol{\epsilon}_S(t) \equiv \mathbf{x}^*(t) - \mathbf{x}_S(t)$, $t \in I$, we have

$$D_t\boldsymbol{\epsilon}_S(t) = A(t)\boldsymbol{\epsilon}_S(t) + B(t)(\mathbf{u}^*(t) - \mathbf{u}_S(t)), \qquad t \in I,$$

and $\boldsymbol{\epsilon}_S(0) = \mathbf{0}$. This implies that

$$\boldsymbol{\epsilon}_S(t) = \int_0^t A(z)\boldsymbol{\epsilon}_S(z)\, dz + \int_0^t B(z)\boldsymbol{\delta}_S(z)\, dz$$

or

$$| \boldsymbol{\epsilon}_S(t) |_2 \leq \int_0^t | A(z) |_2 | \boldsymbol{\epsilon}_S(z) |_2\, dz + \int_0^t | B(z) |_2 | \boldsymbol{\delta}_S(z) |_2\, dz.$$

Applying the Gronwall Inequality (cf. [10.5]), to this last inequality, we obtain

$$| \boldsymbol{\epsilon}_S(t) |_2 \leq \| B \|_2 \| \boldsymbol{\delta}_S \|_2 e^{\int_0^1 |A(z)|_2 dz}$$

and

$$\begin{aligned} \|\boldsymbol{\epsilon}_S\|_2^2 &\equiv \int_0^1 |\boldsymbol{\epsilon}_S(t)|_2^2\,dt \le \|B\|_2^2\,\|\boldsymbol{\delta}_S\|_2^2 e^{2\int_0^1 |A(z)|_2 dz} \\ &\equiv \boldsymbol{\Gamma}^2\,\|\mathbf{u}^* - \mathbf{u}_S\|_2^2, \end{aligned}$$

which, when combined with (10.22), proves (10.23). Moreover, we have

$$|D_t\boldsymbol{\epsilon}_S(t)|_2 \le |A(t)|_2\,|\boldsymbol{\epsilon}_S(t)|_2 + |B(t)|_2|\,\mathbf{u}^*(t) - \mathbf{u}_S(t)|_2, \qquad t \in I,$$

and hence by the triangle inequality

$$\begin{aligned} \|D_t\boldsymbol{\epsilon}_S\|_2 &\le \|A\|_\infty\,\|\boldsymbol{\epsilon}_S\|_2 + \|B\|_\infty\,\|\mathbf{u}^* - \mathbf{u}_S\|_2 \\ &\le (\boldsymbol{\Gamma}\|A\|_\infty + \|B\|_\infty)\,\|\mathbf{u}^* - \mathbf{u}_S\|_2. \end{aligned}$$

Inequality (10.24) follows by using (10.22) to bound $\|\mathbf{u}^* - \mathbf{u}_S\|_2$. Q.E.D.

We now prove a result which gives us an error bound for the cost criteria, i.e., if we actually use the computed control $\mathbf{u}_S(t)$ and the system behaves according to $\mathbf{x}_S(t)$, how does $J[\mathbf{u}_S, \mathbf{x}_S]$ compare with $J[\mathbf{u}^*, \mathbf{x}^*]$? The proof is essentially the same as the one for the analogous result in [10.4].

THEOREM 10.6

Under the preceding hypotheses,

$$\begin{aligned} &J[\mathbf{u}^*, \mathbf{x}^*] \\ (10.25)\qquad &\le J[\mathbf{u}_S, \mathbf{x}_S] \le J[\mathbf{u}^*, \mathbf{x}^*] + \|R^{-1}B^T\|_\infty^2 \xi_Q^{-1}\rho^2(\|Q\|_\infty \boldsymbol{\Gamma}^2 + \|R\|_\infty) \\ &\times \inf_{\mathbf{w}\in S} |\boldsymbol{\lambda} - \mathbf{w}|^2, \end{aligned}$$

where $\boldsymbol{\Gamma}$ is defined in Theorem 10.5 and ξ_Q and ρ are defined in Theorem 10.1.

Proof. If $\boldsymbol{\delta}_S(t) \equiv \mathbf{u}^*(t) - \mathbf{u}_S(t)$, $t \in I$, and $\boldsymbol{\epsilon}_S(t) \equiv \mathbf{x}^*(t) - \mathbf{x}_S(t)$, $t \in I$, then

$$\begin{aligned} J[\mathbf{u}_S, \mathbf{x}_S] &= \tfrac{1}{2}\int_0^1 (\mathbf{x}_S(t), Q(t)\mathbf{x}_S(t))\,dt + \tfrac{1}{2}\int_0^1 (\mathbf{u}_S(t), R(t)\mathbf{u}_S(t))\,dt \\ &= \tfrac{1}{2}\int_0^1 (\mathbf{x}^* + \boldsymbol{\epsilon}_S, Q(\mathbf{x}^* + \boldsymbol{\epsilon}_S))\,dt + \tfrac{1}{2}\int_0^1 (\mathbf{u}^* + \boldsymbol{\delta}_S, R(\mathbf{u}^* + \boldsymbol{\delta}_S))\,dt \\ &= J[\mathbf{u}^*, \mathbf{x}^*] + \int_0^1 (\boldsymbol{\delta}_S, R\mathbf{u}^*)\,dt + \int_0^1 (\boldsymbol{\epsilon}_S\; Q\mathbf{x}^*)\,dt \\ &\quad + \tfrac{1}{2}\int_0^1 (\boldsymbol{\delta}_S, R\boldsymbol{\delta}_S)\,dt + \tfrac{1}{2}\int_0^1 (\boldsymbol{\epsilon}_S, Q\boldsymbol{\epsilon}_S)\,dt. \end{aligned}$$

But since (10.7) must hold for the optimal $\boldsymbol{\lambda}^*$ and $\mathbf{u}^*$, we have

$$\begin{aligned}\int_0^1 (\boldsymbol{\delta}_S, R\mathbf{u}^*)\,dt &= -\int_0^1 (\boldsymbol{\delta}_S, B^T\boldsymbol{\lambda}^*)\,dt \\ &= -\int_0^1 (B\boldsymbol{\delta}_S, \boldsymbol{\lambda}^*)\,dt.\end{aligned} \tag{10.26}$$

However, from the equation (10.2) we have that

$$D_t\boldsymbol{\epsilon}_S(t) = A(t)\boldsymbol{\epsilon}_S(t) + B(t)\boldsymbol{\delta}_S(t),$$

and combining this with (10.26) we obtain

$$\int_0^1 (\boldsymbol{\delta}_S, R\mathbf{u}^*)\,dt = -\int_0^1 (D_t\boldsymbol{\epsilon}_S - A\boldsymbol{\epsilon}_S, \boldsymbol{\lambda}^*)\,dt. \tag{10.27}$$

Integrating the right-hand side of (10.27) by parts, using the boundary conditions on $\boldsymbol{\epsilon}_S$ and $\boldsymbol{\lambda}^*$, and using (10.8) for $\boldsymbol{\lambda}^*$ and $\mathbf{x}^*$, we obtain

$$\begin{aligned}\int_0^1 (\boldsymbol{\delta}_S, R\mathbf{u}^*)\,dt &= \int_0^1 \{(\boldsymbol{\epsilon}_S, D_t\boldsymbol{\lambda}) + (\boldsymbol{\epsilon}_S, A^T\boldsymbol{\lambda}^*)\}\,dt \\ &= -\int_0^1 (\boldsymbol{\epsilon}_S, Q\mathbf{x}^*)\,dt.\end{aligned}$$

Finally, using (10.22) and (10.23) we have

$$\begin{aligned}J[\mathbf{u}_S, \mathbf{x}_S] &= J[\mathbf{u}^*, \mathbf{x}^*] + \tfrac{1}{2}\int_0^1 (\boldsymbol{\delta}_S, R\boldsymbol{\delta}_S)\,dt + \tfrac{1}{2}\int_0^1 (\boldsymbol{\epsilon}_S, Q\boldsymbol{\epsilon}_S)\,dt \\ &\leq J[\mathbf{u}^*, \mathbf{x}^*] + \tfrac{1}{2}\| R \|_\infty \| \boldsymbol{\delta}_S \|_2^2 + \tfrac{1}{2}\| Q \|_\infty \| \boldsymbol{\epsilon}_S \|_2^2 \\ &\leq J[\mathbf{u}^*, \mathbf{x}^*] + \tfrac{1}{2}\| Q \|_\infty \boldsymbol{\Gamma}^2 \| R^{-1}B^T \|_\infty^2 (2\xi_Q^{-1})^2\rho^2 \inf_{\mathbf{w}\in S} | \boldsymbol{\lambda}^* - \mathbf{w} |^2 \\ &\quad + \tfrac{1}{2}\| R^{-1}B^T \|_\infty^2 \xi_Q^{-1}\rho^2(\| Q \|_\infty \boldsymbol{\Gamma}^2 + \| R \|_\infty) \inf_{\mathbf{w}\in S} | \boldsymbol{\lambda}^* - \mathbf{w} |^2.\end{aligned}$$

Q.E.D.

We now consider how these general error bounds can be applied to specific examples. As subspaces of $\boldsymbol{\Phi}_0^n$, we consider

$$\tilde{L}_0^n(\boldsymbol{\Delta}) \equiv \left\{\sum_{i=0}^{N} \boldsymbol{\beta}_i l_i(x) \mid \boldsymbol{\beta}_i \in R^n, \quad 0 \leq i \leq N\right\}, \tag{10.28}$$

$$\begin{aligned}\tilde{H}_0^n(\boldsymbol{\Delta}) \equiv \Bigg\{\sum_{i=0}^{N} \boldsymbol{\beta}_i h_i(x) + \sum_{i=0}^{N+1} \boldsymbol{\alpha}_i h_i^1(x) \mid \boldsymbol{\beta}_i \in R^n,& \\ 0 \leq i \leq N, \text{ and } \boldsymbol{\alpha}_i \in R^n, 0 \leq i \leq N+1\Bigg\},&\end{aligned} \tag{10.29}$$

and

$$\tilde{S}_0^n(\Delta) \equiv \left\{ \sum_{i=-3}^{N-3} \boldsymbol{\beta}_i s_i(x) + \boldsymbol{\beta}_{N-2}\tilde{s}_{N-2}(x) + \boldsymbol{\beta}_{N-1}\tilde{s}_{N-1}(x) \,|\, \boldsymbol{\beta}_i \in R^n, \; -3 \le i \le N-1 \right\}. \tag{10.30}$$

It is easily verified that we obtain a block-banded matrix for the linear system (10.6) by using these subspaces and basis functions in the Ritz procedure. Moreover, combining the results of Theorems 2.5, 3.5, and 4.6 with the general results of this chapter, we obtain the following error bounds.

THEOREM 10.7

If $\boldsymbol{\lambda}^* \in \bigtimes_{i=1}^{n} [PC^{2,2}(I)]_i$, then there exists a positive constant, K, such that

$$\|\boldsymbol{\lambda}^* - \boldsymbol{\lambda}_{\tilde{L}_0^n}\|_2 \le Kh, \tag{10.31}$$

$$\|\mathbf{u}^*, \mathbf{u}_{\tilde{L}_0^n}\|_2 \le Kh, \tag{10.32}$$

and

$$J[\mathbf{u}^*, \mathbf{x}^*] \le J[\mathbf{u}_{\tilde{L}_0^n}, \mathbf{x}_{\tilde{L}_0^n}] \le J[\mathbf{u}^*, \mathbf{x}^*] + Kh^2. \tag{10.33}$$

THEOREM 10.8

If $\boldsymbol{\lambda}^* \in \bigtimes_{i=1}^{n} [PC^{4,2}(I)]_i$, then there exists a positive constant, K, such that

$$\|\boldsymbol{\lambda}^* - \boldsymbol{\lambda}_{\tilde{H}_0^n}\|_2 \le Kh^3, \qquad \|\boldsymbol{\lambda}^* - \boldsymbol{\lambda}_{\tilde{S}_0^n}\|_2 \le Kh^3, \tag{10.34}$$

$$\|\mathbf{u}^* - \mathbf{u}_{\tilde{H}_0^n}\|_2 \le Kh^3, \qquad \|\mathbf{u}^* - \mathbf{u}_{\tilde{H}_0^n}\|_2 \le Kh^3, \tag{10.35}$$

$$J[\mathbf{u}^*, \mathbf{x}^*] \le J[\mathbf{u}_{\tilde{H}_0^n}, \mathbf{x}_{\tilde{H}_0^n}] \le J[\mathbf{u}^*, \mathbf{x}^*] + Kh^6, \tag{10.36}$$

and

$$J[\mathbf{u}^*, \mathbf{x}^*] \le J[\mathbf{u}_{\tilde{S}_0^n}, \mathbf{x}_{\tilde{S}_0^n}] \le J[\mathbf{u}^*, \mathbf{x}^*] + Kh^6. \tag{10.37}$$

We have shown that under the appropriate hypotheses, the Ritz procedure produces an approximate control over $\tilde{L}_0^n$ which is second-order accurate with respect to the performance criteria and approximate controls over $\tilde{H}_0^n$ and $\tilde{S}_0^n$, which are sixth-order accurate with respect to the performance criteria. See [10.4] for some sample numerical results.

EXERCISES FOR CHAPTER 10

(10.1) Prove analogues of the results of Section 10.2 for the subspaces of Φ_0^n generated by the spaces $S(2m-1, \Delta, z)$ defined in Exercise 4.12.

(10.2) Obtain bounds for the constant, K, of Theorems 10.7 and 10.8.

REFERENCES FOR CHAPTER 10

[10.1] ATHANS, M., and P. L. FALB, *Optimal Control: An Introduction to the Theory and Its Applications.* McGraw-Hill, New York (1966).

[10.2] BOSARGE, W. E., and O. G. JOHNSON, Error bounds of high-order accuracy for the state regulator problem via piecewise polynomial approximations. *SIAM J. on Control* **9**, 15–28 (1971).

[10.3] BOSARGE, W. E., and O. G. JOHNSON, Direct method approximation to the state regulator control problem using a Ritz–Treffitz suboptimal control. Proc. Joint Automatic Control Conference, 1970.

[10.4] BOSARGE, W. E., and O. G. JOHNSON, Numerical properties of the Ritz–Trefftz algorithm for optimal control. *Communications of the ACM* **14**, 402–406 (1971).

[10.5] BRAUER, F., and J. NOHEL, *Ordinary Differential Equations.* W.A. Benjamin, Inc., New York (1966).

[10.6] SCHULTZ, M. H., *A Ritz method for an optimal control problem.* Yale Computer Science Research Report 71–9.

INDEX

A

Approximation
  backward difference, 135
  Crank–Nicholson, 135, 138
  forward difference, 135
  Galerkin, 100, 112, 118, 131, 132, 134, 136
  Rayleigh–Ritz, 90, 98, 118
  Tchebyscheff, 78

B

Backward difference approximation, 135
Boot strap method, 33

C

Cauchy–Schwarz Inequality, 7, 17, 34, 54
Cholesky decomposition, 77–78, 101
Crank–Nicholson approximation, 135, 138

D

Decomposition, Cholesky, 77–78, 101
Degenerate kernels, 64
Dirichlet boundary conditions, 88, 93, 99
Dirichlet problem, 87

E

Eigenfunction, 116, 117, 121–122, 123, 124
  approximate, 118
Eigenvalue, approximate, 118
Eigenvalue problems, Rayleigh–Ritz–Galerkin procedure for, 116–126
  error analysis, 120–124
  exercises, 124–125
  one-dimensional problems, 116–120
Eigenvectors, 118
Elliptic differential equations, Rayleigh–Ritz–Galerkin procedure for, 87–115
  error analysis, 102–110
  exercises, 110–112
  linear second-order two-point boundary value problems, 88–93
  second-order problems in the plane, 99–102
  semilinear second-order two-point boundary value problems, 93–99
Error analysis
  finite-element regression, 78–84

Error analysis (*cont.*)
piecewise cubic hermite interpolation, 31–39
piecewise linear interpolation, 14–21
Rayleigh–Ritz–Galerkin procedure
for eigenvalue problems, 120–124
for elliptic differential equations, 102–110
spline interpolation, 51–60
Error bounds, Ritz procedure for an optimal control problem, 145–150
Euler equation, 15, 31–32, 52, 143

F

Finite element regression, 69–86
error analysis, 78–84
exercises, 84–85
one-dimensional problems, 69–75
two-dimensional problems, 76–78
First Integral Relation
piecewise cubic hermite interpolation, 32, 33, 42
piecewise linear interpolation, 16
spline interpolation, 52, 62
Forward difference approximation, 135

G

Galerkin approximation, 100, 112, 118, 131, 132, 134, 136
Galerkin procedures, 90, 97–98, 118
for parabolic equations, 127–140
computational considerations, 134–139
exercises, 139–140
linear problems, 127–131
semilinear problems, 131–134
Gaussian elimination, 1, 47, 70, 73, 91, 93
Gauss–Seidel method, 98
Gerschgorin Theorem, 47, 71
Gronwall Inequality, 128, 147

I

Inequality
Cauchy–Schwarz, 7, 17, 34, 54
Gronwall, 128, 147
Inequality (*cont.*)
Jensen's, 9
Rayleigh–Ritz, 4, 15, 16, 33, 37, 53, 79, 91
Schmidt, 37, 78, 84, 92
Interpolation
Lagrange, 10, 47
piecewise cubic hermite, 24–43
error analysis, 31–39
exercises, 39–43
First Integral Relation, 32, 33, 42
one-dimensional problems, 24–29
Second Integral Relation, 32, 34, 41, 42
spline interpolation and, 44
two-dimensional problems, 29–31
piecewise linear, 10–23
error analysis, 14–21
exercises, 21–23
First Integral Relation, 16
one-dimensional problems, 10–13
Second Integral Relation, 16
two-dimensional problems, 13–14
spline, 44–63
error analysis, 51–60
exercises, 61–63
First Integral Relation, 52, 62
one-dimensional problems, 44–49
piecewise cubic hermite interpolation and, 44
Second Integral Relation, 53, 54, 62
two-dimensional problems, 49–51

J

Jensen's Inequality, 9

K

Kronecker delta function, 11
Kronecker product, 77

L

Lagrange interpolation, 10, 47
Lagrange multiplier, 143
Legendre polynomials, 7
Linear integral equations, 64–68
Linear second-order two-point boundary value problems, 88–93

M

Mapping, orthogonal projection, 120–121

N

Newton's method, 98
Nonzero diagonals, 76

O

One-dimensional problems
finite element regression, 69–75
piecewise cubic hermite interpolation, 24–29
piecewise linear interpolation, 10–13
Rayleigh–Ritz–Galerkin procedure for eigenvalue problems, 116–120
spline interpolation, 44–49
Optimal control problem, Ritz procedure for, 141–151
error bounds, 145–150
exercises, 150
formulation of the procedure, 141–145
Orthogonal projection mapping, 120–121

P

Padé table, 135–136
Parabolic equations, Galerkin procedures for, 127–140
computational considerations, 134–139
exercises, 139–140
linear problems, 127–131
semilinear problems, 131–134
Peano Kernel Theorem, 18, 29, 35, 42, 55, 58, 61
Piecewise cubic hermite interpolation, 24–43
error analysis, 31–39
exercises, 39–43
First Integral Relation, 32, 33, 42
one-dimensional problems, 24–29
Second Integral Relation, 32, 34, 41, 42
spline interpolation and, 44
two-dimensional problems, 29–31
Piecewise linear interpolation, 10–23
error analysis, 14–21
Piecewise linear interpolation (*cont.*)
exercises, 21–23
First Integral Relation, 16
one-dimensional problems, 10–13
Second Integral Relation, 16
two-dimensional problems, 13–14
Plane, second-order problems in, 99–102

R

Rayleigh quotient, 8, 117
Rayleigh–Ritz approximation, 90, 98, 118
Rayleigh–Ritz Inequality, 4, 15, 16, 33, 37, 53, 79, 91
Rayleigh–Ritz procedure, 89, 95, 100, 117
Rayleigh–Ritz–Galerkin procedure
for eigenvalue problems, 116–126
error analysis, 120–124
exercises, 124–125
one-dimensional problems, 116–120
for elliptic differential equations, 87–115
error analysis, 102–110
exercises, 110–112
linear second-order two-point boundary value problems, 88–93
second-order problems in the plane, 99–102
semilinear second-order two-point boundary value problems, 93–99
Regression, finite element, 69–86
error analysis, 78–84
exercises, 84–85
one-dimensional problems, 69–75
two-dimensional problems, 76–78
Ritz procedure for an optimal control problem, 141–151
error bounds, 145–150
exercises, 150
formulation of the procedure, 141–145
Rolle's Theorem, 4, 7, 18, 35, 53

S

Schmidt Inequalities, 37, 38, 84, 92
Second Integral Relation
piecewise cubic hermite interpolation, 32, 34, 41, 42
piecewise linear interpolation, 16

Second Integral Relation (*cont.*)
spline interpolation, 53, 54, 62
Second-order problems in the plane, 99–102
Semilinear second-order two-point boundary value problems, 93–99
Simpon's rule, 57
Spline interpolation, 44–63
error analysis, 51–60
exercises, 61–63
First Integral Relation, 52, 62
one-dimensional problems, 44–49
piecewise cubic hermite interpolation and, 44
Second Integral Relation, 53, 54, 62
two-dimensional problems, 49–51

T

Taylor series, 97
Taylor's Theorem, 6
Tchebyscheff approximation, 78
Two-dimensional problems
finite element regression, 76–78
piecewise cubic hermite interpolation, 29–31
piecewise linear interpolation, 13–14
spline interpolation, 49–51

U

Uniform partitions, 29